The Failures of Mathematical Anti-Evolutionism

Anti-scientific misinformation has become a serious problem on many fronts, including vaccinations and climate change. One of these fronts is the persistence of anti-evolutionism, which has recently been given a superficially professional gloss in the form of the intelligent design movement. Far from solely being of interest to researchers in biology, anti-evolutionism must be recognized as part of a broader campaign with a conservative religious and political agenda. Much of the rhetorical effectiveness of anti-evolutionism comes from its reliance on seemingly precise mathematical arguments. This book, the first of its kind to be written by a mathematician, discusses and refutes these arguments. Along the way, it also clarifies common misconceptions about both biology and mathematics. Both lay audiences and professionals will find the book to be accessible and informative.

JASON ROSENHOUSE is a professor of mathematics at James Madison University. He is the author or editor of eight previous books, including *Among the Creationists: Dispatches from the Anti-Evolutionist Front Line* (Oxford University Press, 2012).

"If you want to convince an audience of a falsehood, an excellent strategy is to bamboozle them with mathematical jargon that they don't understand. If your incomprehensible equations seem to prove what the audience wants to believe anyway, you'll earn a standing ovation. This has been a favorite tactic of anti-evolutionists. Jason Rosenhouse is a good mathematician but, unlike many mathematicians, he is superbly gifted in the art of translating mathematical arguments into words. He mercilessly exposes creationist abuses of mathematics, using language that non-mathematicians can pleasurably follow. As a bonus along the way, he is an excellent mathematics teacher, patiently explaining each point before revealing the abuse of it. I especially appreciated his deft use of analogies. He is thoroughly well read in the bogus literature of creationism and 'intelligent design,' as well as in the biological literature. We have here a superb book, lucid, knowledgeable, wise, and very necessary."

Richard Dawkins FRS, Emeritus Professor of the Public Understanding of Science, University of Oxford

"A little mathematics can be a dangerous thing when it is used as a rhetorical weapon in a political and religious battle. In this incisive and crystal clear book, Jason Rosenhouse shows that the arguments of mathy creationists are like the proverbial spherical cow and unflightworthy bumblebee: sophistical, wrongheaded, and out of touch with biological reality."

Steven Pinker, Johnstone Professor of Psychology, Harvard University, and the author of *How the Mind Works* and *Rationality*

"Creationists beware! Jason Rosenhouse is here to debunk your perversions of probability theory, your tortured thermodynamics, and your insults to information theory. And the rest of us will learn a lot about evolution, mathematics, and their interrelations from the expert guidance offered in *The Failures of Mathematical Anti-Evolutionism.*"

Glenn Branch, Deputy Director, National Center for Science Education

"Written by a mathematician who has paid close attention to mathematical arguments against evolution, this is a beautifully written and careful refutation of those arguments. Jason Rosenhouse covers arguments based on improbability, information, and thermodynamics, and he does so with clear explanations that will be accessible to anyone who takes science seriously. A wonderful achievement."

Joe Felsenstein, University of Washington

The Failures of Mathematical Anti-Evolutionism

JASON ROSENHOUSE
James Madison University

CAMBRIDGE
UNIVERSITY PRESS

CAMBRIDGE
UNIVERSITY PRESS

University Printing House, Cambridge CB2 8BS, United Kingdom

One Liberty Plaza, 20th Floor, New York, NY 10006, USA

477 Williamstown Road, Port Melbourne, VIC 3207, Australia

314–321, 3rd Floor, Plot 3, Splendor Forum, Jasola District Centre,
New Delhi – 110025, India

103 Penang Road, #05–06/07, Visioncrest Commercial, Singapore 238467

Cambridge University Press is part of the University of Cambridge.

It furthers the University's mission by disseminating knowledge in the pursuit of
education, learning, and research at the highest international levels of excellence.

www.cambridge.org
Information on this title: www.cambridge.org/9781108842303
DOI: 10.1017/9781108907149

First published 2022

A catalogue record for this publication is available from the British Library.

Library of Congress Cataloging-in-Publication Data
Names: Rosenhouse, Jason, author.
Title: The failures of mathematical anti-evolutionism / Jason Rosenhouse.
Description: New York : Cambridge University Press, 2021. |
    Includes bibliographical references and index.
Identifiers: LCCN 2021041875 | ISBN 9781108842303 (hardback) |
    ISBN 9781108820448 (paperback) | ISBN 9781108907149 (epub)
Subjects: LCSH: Evolution (Biology) – Mathematics. |
    Evolution (Biology) – Mathematical models.
Classification: LCC QH371.3.M37 R67 2021 | DDC 576.801/5118–dc23
LC record available at https://lccn.loc.gov/2021041875

ISBN 978-1-108-84230-3 Hardback
ISBN 978-1-108-82044-8 Paperback

To my friends at the National Center for Science Education.
Thanks for everything you do.

Contents

Preface

People who work in the life sciences typically regard it as obvious that modern evolutionary theory is essentially correct. They find that there is just too much data that falls right into place if you take evolution as your starting point, and they consistently get good results when they apply the theory to practical problems. There is much to argue about in the details, and new ideas seem to get introduced faster than they can be assessed, but there is near unanimous agreement that the big picture – that modern species are the end result of a lengthy historical process and that natural selection is an especially important mechanism of evolution – is roughly the way Charles Darwin first described it in 1859.

But for as long as there have been evolutionists there have also been anti-evolutionists. There have always been those who *do not like* evolution, and they offer a variety of arguments in support of their views. Many of these arguments are at least superficially scientific, though it is hardly a secret that religious motivations are nearly always lurking beneath the surface. For their part, scientists mostly scoff at anti-evolutionism, and rightly so. Some anti-evolutionists write more compellingly than others, but it never really seems too difficult for a scientifically knowledgeable reader to refute their arguments. Frankly, most anti-evolutionist arguments are based on faulty reasoning, absurd distortions of the scientific facts, caricatures of what evolutionary theory actually asserts, or all three.

Mathematics has long played a role in anti-evolution discourse, but in the past 10–15 years it has become far more prominent than it had previously been. Nearly all of the major anti-evolution books and articles during this time place mathematical arguments front and center. Sometimes it seems like every other page contains a probability calculation, an invocation of information theory, or a dubious claim about combinatorial search. Scientists have replied piecemeal to these arguments, but there has been no survey of mathematical anti-evolutionism taken as a whole. Moreover, it is only on rare occasions that mathematicians themselves have taken note of this abuse of their discipline. Our training makes us sensitive to points that are likely to be overlooked by scientists, for whom mathematics is mostly just a tool that they use in doing their work. That is where this book comes in.

As I describe in Section 1.3, my interest in this subject began in the early 2000s. I had just completed my PhD in mathematics, and my first job out of graduate school was at Kansas State University. Specifically, my job had much to do with the training of public school mathematics teachers in the state. At that time, Kansas was mired in controversy because a politically conservative state school board had voted to remove scientific topics like evolution and the big bang from the science curriculum. My work brought me into close contact with people on the front lines of this dispute. When I subsequently learned of a forthcoming creationism conference not far from my home, I decided, on a whim, to attend.

Over the next 7–8 years I attended dozens of such conferences, as well as smaller gatherings. Some were devoted to young-Earth creationism, which holds, among other things, that the best available science confirms a literal reading of the creation accounts in the early chapters of the biblical book of Genesis. Others promoted intelligent design, a more modest form of creationism that came to prominence in the early 1990s. Some were large conferences like the one I attended in Kansas, while others were much smaller gatherings held in local churches.

After many years of this I decided I had a story to tell, and I told it in my earlier book, *Among the Creationists: Dispatches from the Anti-Evolutionist Front Line*, published in 2011. The book was mostly about the cultural milieu in which these disputes play out, and it was organized around my personal experiences. It was not primarily about refuting creationist arguments, though I did address a few of them along the way.

I had intended for that to be my last word on the subject, but the strong emphasis on mathematical arguments in recent anti-evolutionist discourse made me reconsider. I thought about all the times I had seen anti-evolutionists present transparently fallacious mathematical arguments to their audiences, only to be rewarded with cheers and standing ovations as a result. I also thought about all the times when I had seen scientists respond to anti-evolutionist arguments in ways that I thought did not really get to the heart of the matter. It is for these reasons that I decided to write this book.

The first three chapters are mostly stage setting. After providing a general introduction in Chapter 1, I present some of the basics of evolutionary biology and mathematics in Chapters 2 and 3. The evidence for evolution has been presented at length in other venues, but since I wanted this book to be as self-contained as possible, I included some of that material in Chapter 2. Many people hold misconceptions about what mathematics really is since they tend to equate it with arithmetic. I dispel those myths in Chapter 3.

Chapter 4 addresses the famous Wistar conference, held in Philadelphia in 1966. The proceedings were published the following year under the title *Mathematical Challenges to the Neo-Darwinian Theory of Evolution*. The challenges came primarily from Murray Eden and Marcel-Paul Schützenberger, respectively an engineer at the Massachusetts Institute of Technology and a mathematician at the University of Paris. Eden and Schützenberger were both very eminent in their own fields, but their anti-evolution arguments were just bad. Their ideas cut little ice with the assembled biologists, but they remain folk heroes among anti-evolutionists to this day. We will analyze their arguments in considerable detail.

Chapters 5–7 constitute the heart of the book. We consider the major lines of argument in modern mathematical anti-evolutionism, and we explain why they are entirely misguided. It is not only that the specific arguments they make contain errors of various sorts, but also that their whole way of thinking about evolution is fundamentally mistaken. Chapter 5 discusses arguments based on probability theory, while Chapter 6 considers arguments drawn from the closely related mathematical theories of information and combinatorial search.

Chapter 7 addresses anti-evolution arguments based on thermodynamics, especially the second law. Thermodynamics would normally be considered a branch of physics, but it has a strongly mathematical character that justifies its inclusion in this book. The second law is a precise, mathematical statement, and a failure to appreciate this fact allows anti-evolutionists to get away with incredibly sloppy argumentation, and it leads evolutionists to not always reply as effectively as they might do. We then close with a short epilogue (Chapter 8), summarizing the discussion and offering a few closing thoughts.

Let me be clear that this is a mathematics book that also discusses biology, as opposed to a biology book that also discusses mathematics. Inevitably, there are places where we must get our hands dirty by digging into the biological details, but my central points are mathematical and not biological. Biologists will rightly criticize me for presenting a simplistic version of modern evolutionary theory. I focus almost entirely on natural selection acting at the level of genes, but everyone understands that there is far more to evolution than this. The geneticists will likewise have reason to complain. In this book, a gene is treated as nothing more than a combinatorial sequence drawn from an alphabet of four letters, thereby ignoring most of the difficult technical details about how genes actually work.

My excuse is that the anti-evolutionists under consideration aim their fire almost exclusively at the question of how complex adaptions arise, and this justifies my narrow focus. I am not writing a general treatise on evolutionary biology. Instead, I am discussing the one small part of the theory relevant to anti-evolutionist arguments. Moreover, I am rhetorically making things more difficult for evolution by restricting its explanatory options solely to natural selection acting on chance genetic variations. My argument is essentially that even if we take this narrow understanding of evolution as our starting point, we still have more than enough resources to refute any gambit coming from the other side.

My intended audience is anyone who takes an interest in the evolution/creation issue. I have tried to write in as nontechnical a manner as possible, and I have mostly avoided notation and jargon. The handful of places where I did include notation can be skimmed without losing the flow of the argument. For lay audiences, I hope I have managed to provide some food for thought about mathematics, and that I have shown that scientists have good reasons for being dismissive of mathematical anti-evolutionism. For professional scientists, the perspective of a mathematician on these issues might hold some interest.

In the end, this book was written in the conviction that nonsense has to be confronted. The anti-evolutionists are slick and well funded, and their output is just one part of a broader, politically conservative agenda. I have no illusions that one short academic book can really put much of a dent in the anti-scientific parallel universe they have created, but I believe it is important to try.

Several people provided helpful advice and guidance while I was preparing the manuscript. Kostas Kampourakis first suggested that I write this book, and I am very glad he did so. Initially I was skeptical – a whole book on mathematical anti-evolutionism? – but as I immersed myself in the project I came to agree that it was necessary. Glenn Branch, Tom English, Joseph Felsenstein, and Burt Humburg provided invaluable feedback on the first draft of the book. I have greatly benefited from their scientific, historical, and literary expertise, and the final product has been strengthened as a result. Of course, it should go without saying that the fault for any remaining errors lies entirely with me. Finally, let me thank Katrina Halliday, Olivia Boult, Divya Arjunan and their whole team at Cambridge University Press for supporting this project and for being understanding when I shot past my deadlines.

I Scientists and Their Hecklers

I.I DARWIN PRESENTS HIS THEORY

Charles Darwin presented his theory of evolution in his book *On the Origin of Species*, published in 1859. In so doing, he transformed biology from a scientific backwater to a fully professional science. Prior to Darwin, biology was little more than the art of catching an animal, killing it, cutting it open, and then writing detailed descriptions of what you saw. Alternatively, some biologists concerned themselves with classifying organisms according to whatever arbitrary characteristics had their attention that week. Valuable work, no doubt, but hardly a science. *Real* science involved abstract theorizing, mathematical modeling, and predictive power to several places past the decimal point. Or so went the stereotype, at any rate.

That all changed with Darwin. By marshaling evidence from classification, biogeography, embryology, and comparative anatomy, he established, to the satisfaction of most scientists, that organisms shared a far greater degree of relatedness than had previously been appreciated. He also provided a possible mechanism to explain how populations of organisms gradually became better adapted to their environments – the process of natural selection. He anticipated, and provided cogent replies to, numerous theoretical objections to his ideas. Biology now had a bona fide theory from which to work, one which could be tested against data and which suggested fruitful directions for further research.

The ensuing 162 years (keeping in mind that I am writing this in early 2021) led to one success after another for evolution. Realizing that a thorough understanding of heredity was necessary for assessing the theory, scientists undertook a program of research that eventually led to the modern science of genetics. In the 1920s and 1930s, mathematical models were developed to help understand gene flow and other evolutionary processes, thereby showing that natural selection was not just possible but also plausible as a mechanism for large-scale evolution. In the 1940s, developments in paleontology, genetics, physiology, zoology, and botany were united into the so-called modern synthesis, showing that the data from every branch of the life sciences seemed to converge on evolution, with natural selection as its primary mechanism. Subsequent developments in molecular biology, and technological developments that made possible new research directions in genetics, provided lines of evidence for evolution undreamed of by Darwin or his immediate successors. The more that was learned about biology, the more evolution came to seem obvious.

Evolutionary thinking soon led to progress in other branches of science. Ecologists realized that evolution was essential to understanding the temporal and spatial distribution of species. Medical researchers came to use evolutionary thinking to understand the process of antibiotic resistance in bacteria, to investigate the origins of genetic disorders, and to devise effective treatments against a host of ailments. Computer scientists used genetic algorithms to solve problems in engineering, meaning they explored large spaces of possibilities by mimicking the process of evolution by natural selection.

Today, evolutionary theory retains pride of place in biological thinking. Modern evolutionary biology includes a large role for Darwin's main ideas, in the sense that the common descent of all modern organisms is considered to be beyond dispute and natural selection is still considered to be an especially important mechanism of evolution. But the subject has also been enriched by many ideas that go well beyond anything Darwin considered. Research into evolution seems to generate novel ideas faster than they can be assessed and assimilated. The field is marked by ferment over details coupled with confidence in the fundamentals.

However, there are today, and always have been, others who are unimpressed by this long track record of success. For as long as biologists have been studying the processes of biological evolution, there have been critics heckling them from the sidelines. The critics claim that evolution is only weakly supported by the available evidence, to the extent that it is supported at all. They claim that evolution has represented a tragic wrong turn in the history of ideas, and that it must be replaced, or at least heavily supplemented, with the idea that an intelligent designer is in some way manipulating the process. In their more florid moments, they claim that evolution is a flatly ridiculous theory, that nothing more than common sense and a high school education is sufficient to see this, and that scientists are blind to this reality because of morbid anti-religious bias.

They make many arguments in support of this view. Some of those arguments rely heavily on mathematics. This book explains why those mathematical arguments are wrong.

1.2 WHO ARE THE HECKLERS?

In the United States in the twenty-first century, there are two main schools of anti-evolutionist thought: Young-Earth Creationism (YEC) and Intelligent Design (ID). You can certainly identify other schools and draw subtle distinctions among their various religious commitments, but the fact remains that YEC and ID all but monopolize the discourse.

YEC holds that Earth was created no more than 10,000 years ago. (Relative to the more standard scientific estimate of roughly four and a half billion years, this constitutes a young Earth.) YEC also claims that modern species were created in essentially their present form. Moreover, it claims that species can be grouped together into distinct "kinds," and that while small amounts of evolutionary change within a kind are possible, more significant change between kinds is not. The basic facts of geology and paleontology, they go on to argue, are best explained by reference to a global deluge a few 1,000 years ago. Critically, they claim that while these ideas are certainly consonant with what is presented in the early chapters of the biblical book of Genesis, they are nonetheless also supported by our best current understanding of the scientific data.

ID is far more modest. It claims only that natural selection is insufficient to explain certain aspects of modern organisms and that therefore modern evolutionary theory is fundamentally flawed. Proponents of ID also claim they can prove that even in principle no naturalistic mechanism can fully explain the interlocking complexity of modern organisms and that a satisfactory explanation can only be found by appealing to some sort of action on the part of an unspecified intelligent designer. They take no stand on the age of Earth, though most of ID's leading representatives accept that Earth is older than the biblical chronologies suggest. They also have nothing much to say about the identity, abilities, and motivations of the designer, nor do they tell us what the designer actually did. There is really little more to their scientific theorizing than the assertion that an intelligent designer of unspecified motives and abilities did something at some point in natural history.

There are cultural differences between the two groups. Proponents of YEC generally endorse the anti-evolution arguments presented by proponents of ID, but they also find that ID does little to promote religious evangelism. They argue that vague references to an unspecified designer do nothing to win souls for Christ and that this is a serious shortcoming of ID. While they are adamant that their views can be defended entirely on scientific grounds, they also make no secret of their religious motivations.

On the other side, proponents of ID are mostly contemptuous of YEC. They find that YEC literature is generally of such low quality that it brings disrepute to the whole project of anti-evolutionism. The leading proponents of ID are better credentialed than their counterparts in YEC, and they express themselves with far more scientific sophistication than most proponents of YEC can muster.

These differences are real and important. Nonetheless, the proper analogy for the relationship of YEC to ID is that of different dialects of the same language. Both are religiously motivated attacks on evolution, and both camps see the evolution/creation dispute as one front in a larger culture war. While ID proponents are more skillful at deploying scientific jargon, the arguments presented by the two camps are essentially the same.

This leads us to the most important similarity of them all: scientists are all but unanimous in finding both ID and YEC arguments to be entirely fallacious. In most cases, scientists do not even find the arguments interesting or thought provoking. They just find them to be wrong for crass and obvious reasons.

While ID and YEC both have considerable cultural cachet, we will be spending far more time discussing the arguments of the former than the latter. Our interest in this book is solely in the merits of their mathematical arguments as applied to evolution, so we will not give any further consideration to the cultural milieu in which these arguments are presented. The arguments stand or fall on their own merits, independent of any unsavory motivations underlying them.

That acknowledged, it is pointless to deny that certain overly-conservative interpretations of religion are at the foundation of modern anti-evolutionism. And since we are going to conclude that the anti-evolutionist's mathematical arguments are very poor, it is reasonable to keep their unscientific motivations in mind as we consider them.

1.3   BAD MATH CAN BE RHETORICALLY EFFECTIVE

My introduction to anti-evolutionism came a little over 20 years ago when I was a graduate student studying mathematics at Dartmouth College. While I was there, the student newspaper published an opinion piece by a creationist student. In part because I was looking for a distraction from my thesis research, which was not going well at that time, I used it as an opportunity to learn more about the evolution/creation dispute.

Initially, I did not have a strong opinion on this issue one way or the other. I have never been especially religious, and I certainly was not inclined to treat the book of Genesis as a literal, historical account. However, I was open to the possibility that biologists, precisely because they were so often attacked by religious demagogues, had overreacted by exaggerating the strength of their case.

Figuring that I at least knew the basics of evolutionary biology, I started by working my way through a stack of creationist books and articles. What I found was a bewildering array of arguments drawn from numerous branches of science. Creationist authors discussed fossils in one chapter, then genetics in the next, then anatomy, then physics, and on and on. Never having made a serious study of these fields up to that time, I often did not have cogent replies at my fingertips. Still, I was skeptical of the sheer magnitude of their accusations and the extreme simplicity of their arguments. People study for years to become experts in any one of those disciplines, but here was a creationist author with no particular credentials telling me that the professionals in almost every branch of science were just foolish and incompetent. I was expected to believe that the professionals had simply overlooked things that would have been obvious to a bright high school student. That seemed unlikely.

The near-unanimous scientific consensus in support of evolution has held up for well over a century. Now, it is certainly true that entrenched ideas can become so ossified and unquestioned that rival theories find it difficult to get a fair hearing. Just as with every other human enterprise, professional science sometimes confronts its practitioners with social or political pressure to conform to the dominant paradigm. For these reasons, I would never consider the mere fact of consensus to be proof that the theory is correct.

However, I do think a long-standing consensus in support of a theory counts for *something*. To me it suggests that while the theory might be wrong, it is not going to turn out to be crazy. We can always imagine some future discovery that forces us to rethink fundamental ideas, but it is hard to imagine that a well-supported theory will suddenly collapse because a talented amateur notices a conceptual error at the heart of the entire enterprise. If you possess any skeptical impulses at all, then claims of that sort really ought to trigger them.

This skepticism was justified for me by the abuse of mathematics in creationist discourse. Their arguments frequently used probability theory, and they often carried out specific calculations meant to convince me that evolution had been refuted. (We will discuss arguments of this sort in Section 5.5.) The fine points of paleontology and biology might have been beyond me at that time, but I certainly knew a bad probability argument when I saw one. To be clear, I am not speaking now of subtle errors. I am not saying they raised interesting questions, but had overlooked some difficult, technical point. I am talking instead of errors that betrayed an utter incomprehension of the subject.

I reasoned that if creationists were *that* wrong when discussing topics with which I was very familiar, what confidence could I have that their arguments in other branches of science were any more cogent? As I delved into the responses to creationists provided by scientists and philosophers, and more importantly as I had the chance to discuss these questions in person with the relevant professionals, it became clear that I was right to be very skeptical.

I finished graduate school in 2000 and accepted a postdoctoral position (academic speak for an internship) at Kansas State University. A significant portion of my job involved issues in public education, specifically related to the training of mathematics teachers. At that time, Kansas was the focus of national controversy because a politically conservative state school board had voted to eliminate all mention of evolution in the state's standards for science teachers. This put the evolution/creation issue back on my radar, and when I subsequently learned of a large creationist conference taking place near to my home, I decided to attend.

Over the next 8 years or so, both in Kansas and later when I moved to the western part of Virginia, I attended a great many gatherings related to anti-evolutionism. Some were large conferences like the one I attended in Kansas, and others were small, one-day gatherings in local churches. Some of these meetings were devoted to YEC, while others were about ID. Regardless, mathematical arguments were prominent at both. The reactions of the conference goers led me to the conclusion in the title of this section.

For example, at one major creationist conference, I was in the audience for a keynote talk devoted to the branch of mathematics known as "information theory." There were roughly two thousand people in the audience. The speaker went on for close to an hour about how insights from this field could be used to refute evolution and to support creationism. When the talk ended, the audience erupted into a standing ovation. The host of the conference session said, in awe-struck tones, that this was one of the most powerful apologetic arguments he had ever heard. My reaction was considerably more critical. Apparently, where I had seen an absurd caricature of a major branch of mathematics, the audience had seen mathematical support for their religious convictions. (We will look at arguments of this sort in Chapter 6.)

Another time, at a conference promoting ID, I was in a small breakout session of about twenty people. The speaker presented a probability calculation of the sort to which I referred a few paragraphs ago. The result of the calculation was a very small number, and the speaker breathlessly informed the audience that this showed that evolution required us to believe that something extremely improbable, if not flatly impossible, had occurred. At the end of the talk, an audience member said, with a facial expression that suggested the utmost seriousness, "When scientists are confronted with a number that small," and here he paused for dramatic effect, "what else can they do but just stare at it helplessly?" Many of the other audience members offered vigorous nods in response. When it was my turn to speak, I suggested that an alternative to staring helplessly was to question the assumptions underlying the calculation, and I pointed to several ways that those assumptions were hopelessly unrealistic. The audience was not amused.

I could provide many further anecdotes of this sort. Mathematics is unique in its ability to bamboozle a lay audience, making it well suited to the cynical machinations of anti-evolutionist speakers and authors. As a mathematician, I take some offense at that. In large measure, that is why I decided to write this book.

1.4   DOES EVOLUTION HAVE A MATH PROBLEM?

Though Darwin was largely successful at persuading scientists of the fact of common descent, he also faced formidable critics. In the later decades of the nineteenth century, it was still possible to be a scientifically informed skeptic of evolution, especially of the idea that natural selection was a plausible mechanism for large-scale change. First-rate scientists like Louis Agassiz and St. George Mivart placed themselves in opposition to Darwin's ideas, and their arguments could hardly be dismissed as the ignorant ravings of religious demagogues. For his part, Darwin offered forceful replies to the critics, and the debate petered out to something of a draw. Darwin presented a strong case for common descent and a decent plausibility argument for natural selection, but there were numerous gaps that could only be filled by further research.

By the early twentieth century, the debate landscape had changed in at least two ways. Scientifically, the case for evolution only became stronger. Paleontologists found numerous transitional fossils that made it easier to accept the possibility of large-scale transmutation in the course of natural history. Advances in the study of heredity showed that the proposed rivals of natural selection were not workable, and mathematical modeling established that selection could be a more powerful force than had been previously understood. These and other research findings were all consistent with the main ideas of evolutionary theory, and this made it harder to be an informed critic.

Meanwhile, evolution had made the jump from an esoteric theory of interest primarily to professional scientists to an idea that pervaded the culture more generally. The theory made its way into public school curricula, and religious fundamentalists saw this as nothing less than an attack on the souls of their children.

These two shifts – the growing strength of the scientific case for evolution coupled with its increased cultural presence – led to a dramatic decline in the quality of anti-evolutionist discourse. Where once the critics could boast of giants like Agassiz and Mivart, now their most visible advocates were amateur scientists like George McCready Price and politicians like William Jennings Bryan. Cogent scientific arguments against evolution became more difficult to find, but imprecations against godless scientists and creeping materialism were commonplace. This sort of advocacy came to a head in the events of the Scopes "monkey" trial in Tennessee in 1925. Culturally, the legacy of the trial was that anti-evolutionism became all but synonymous with an especially obscurantist form of religion.

As representative of the poor state of their argumentation, let us consider a small book by William A. Williams called, *The Evolution of Man Scientifically Disproved, in 50 Arguments*, the final version of which was published in 1928. Williams was a Presbyterian clergyman, and he placed mathematical arguments front and center in his argumentation. He writes,

> Every theory to which mathematics can be applied will be proved or disproved by this acid test. Figures will not lie, and mathematics will not lie even at the demand of liars. Their testimony is as clear as the mind of God. ... The evolution theory, especially as applied to man, likewise is disproved by mathematics. The proof is overwhelming and decisive. Thus God makes the noble science of mathematics bear testimony in favor of the true theories and against the false theories.
>
> *(Williams 1928, 3–4)*

Williams helpfully numbered and labeled his arguments, so let us see two examples of what he regarded as overwhelming and decisive proofs.

Argument 1 is called "The Population of the World." The thrust of the argument is that the human population is too small, if we believe that humanity has existed in excess of 100,000 years, as evolution would seem to require. He cites census data from 1922 to put the human population of the world at 1,804,187,000. If we imagine starting with a single human pair, which then doubles to 4 people, then 8, and so on, then it is a routine calculation to show that between 30 and 31 doublings are necessary to reach a population of 1.8 billion. He then carries out some calculations to show that if we assume the biblical chronology to be correct, which, he says, places the human population at two 5,177 years previously, then we conclude humanity doubles its population every 168.3 years. He carries out a separate calculation to arrive at the conclusion that the Jewish people have doubled their numbers every 161.251 years, and he makes much ado of the closeness of these numbers.

This is a prelude to the argument's climax, which goes like this:

> [L]et us suppose that man, the dominant species, originated from a single pair, only 100,000 years ago, the shortest period suggested by any evolutionist (and much too short for evolution) and that the population doubled in 1612.51 years, one-tenth the Jewish rate of net increase, a most generous estimate. The present population of the globe should be 4,660,210,253,138,204,300 .... In these calculations, we have made greater allowances than any self-respecting evolutionist could ask without blushing. And yet, withal, it is as clear as the light of day that the ancestors of man could not possibly have lived 2,000,000 or 1,000,000 or 100,000 years ago, or even 10,000 years ago; for if the population had increased at the Jewish rate for 10,000 years, it would be more than two billion times as great as it is. No guess that was ever made, or ever can be made, much in excess of 5177 years, can possibly stand as the age of man. The evolutionist cannot sidestep this argument by a new guess. QED.   (Williams 1928, 11)

To clarify, if humanity originated as a single pair 100,000 years ago and doubled its numbers every 1,612.51 years, then that makes just over 62 doublings. If you raise 2 to the 62 power, the result is in the neighborhood of 4.6 quintillion, as Williams asserts.

Williams is quite taken with this sort of thing, and he develops his full argument over nearly four full pages. I have chosen the one paragraph above as representative both of the argument itself and the writing style Williams employed.

However, I am sure that in the time it took you to read that paragraph, you noticed that Williams based his calculations on a highly dubious assumption. Specifically, he assumed that the human doubling time has been constant throughout our history as a species, but this is not reasonable. Modern scientific estimates suggest that species *Homo sapiens* first appeared roughly 200,000 years ago, but for most of those years the human population was either flat or even decreasing. After all, for most of human history life was nasty, brutish, and short, to use Thomas Hobbes' memorable phrase. Even well into the modern era there have been periods of declining human population, resulting from plagues and famines, for example. It is only with relatively modern advancements in medicine and nutrition that human populations double their numbers with any sort of alacrity. Once you dispense with the assumption of a constant doubling time, Williams' argument comes to look a bit silly.

Let us try one more. Argument 9 is called "Mathematical Probability." Here are some representative quotations:

> The evolution of species violates the rule of mathematical probability. It is so improbable that one and only one species out of 3,000,000 should develop into man, that it certainly was not the case. All had the same start, many had similar environments. ... While all had the same start, only one species out of 3,000,000 reached the physical and intellectual and moral status of man. Why only one? Why do we not find beings equal or similar to man, developed from the cunning fox, the faithful dog, the innocent sheep, or the hog, one of the most social of all animals. Or still more from the many species of the talented monkey family? Out of 3,000,000 chances, is it not likely that more than one species would attain the status of man?
>
> *(Williams 1928, 23–24)*

He does not explain why he thinks it is unlikely that human-like intelligence would evolve only once, but we get a clue to his reasoning from this:

> Evolution is not universally true in any sense of the term. Why are not fishes *now* changing into amphibians, amphibians into reptiles, reptiles into birds and mammals, and monkeys into man? If growth, development, evolution, were the rule, there would be no lower order of animals for all have had sufficient time to develop into the highest orders. Many have remained the same; some have deteriorated. *(Williams 1928, 25)*

After several pages of this, Williams presents his conclusion:

> To declare that our species alone crossed this measureless gulf, while our nearest relatives have not even made a fair start, is an affront to the intelligence of the thoughtful student. It does fierce violence to the doctrine of mathematical probability. It could not have happened. *(Williams 1928, 27)*

Those of you possessing some basic familiarity with evolutionary thinking will be scratching your head at this, since it is hard to understand what Williams is going on about. Every species possesses *some* attribute that makes it unique in the world. Williams could as easily have wondered why the giraffes alone have evolved such excessively long necks, or why it was just a few species of elephants that evolved excessively long noses. The long-term trajectory of evolution is governed by so many variables that it is impossible to predict which life forms and which adaptations will actually appear after millions of years.

The real action in Williams' argument seems to be in that middle quote, where he strongly implies that evolution entails a steady progression from lower to higher forms of life. This is a serious misapprehension, albeit a fairly common one. There is no concept of "higher" or "lower" animals in evolution, and there is certainly no notion that species are striving to ascend some ladder of progress.

We humans tend to be rather self impressed, and we naturally find it tempting to place ourselves at the top of creation. However, evolution only cares about brute survival. A successful animal is one that inserts many copies of its genes into the next generation, and one can do *that* while being not very bright at all.

We should also take note of Williams' casual references to the "rule of mathematical probability" and later to the "doctrine of mathematical probability." There are many available textbooks on probability theory (a statement that would have been no less true in 1928), but you will search them in vain for any reference to a central rule or doctrine at the heart of the subject. From the context, Williams seems to envision a bland statement to the effect that extremely improbable things do not occur, but even this would need a lot of caveats to be credible, since highly improbable things occur all the time. (There is an old saying that in New York City, which has a population of more than eight million people, million to one odds happen eight times a day.)

Williams' argument does bring up a number of interesting questions. For example, we might ask about the engineering constraints and selection pressures that determine whether or not human-like intelligence can evolve. Or we might remark on the phenomenon of evolutionary convergence, in which very similar adaptations arise in separate lineages independently of one another. To discuss those questions here, however, would be to give Williams' book more respect than it deserves.

Instead we should remark on the smug, arrogant tone of his writing, as well as the entirely unjustified bravado. Williams drapes his population calculations over many pages, but, as we have noted, his argument is killed stone dead by the utterly trivial observation that the human doubling rate is not constant over time. How could he not have noticed that? His argument about probability betrayed a complete ignorance of fundamental issues in both biology and mathematics. This is not the work of a man who has taken the views of his opponents seriously, who has devoted some time to understanding the scientific and mathematical concepts about which he is writing, or who has tried to express himself with care and cogency.

This makes him entirely typical among anti-evolutionist writers.

1.5   THE SEARCH FOR AN IN-PRINCIPLE ARGUMENT

Mathematical anti-evolutionism is very ambitious in that it tries to rule evolution out of bounds in principle. If this approach succeeds, then all the circumstantial evidence in the world will be insufficient to save the theory.

To see what I mean, imagine that you are an attorney representing a client accused of murder. The prosecution has multiple pieces of evidence against your client: His fingerprints were found at the scene; he had access to the murder weapon; he had a strong motive; and a person matching his description was seen in the area at the time of the crime. You are trying to devise a defense strategy in response. There are two general approaches you might pursue.

The first approach is to challenge each piece of evidence individually: Your client had an entirely innocent reason for being at the scene prior to the crime, and that is why his fingerprints were found there; many people had a motive and access to the murder weapon; the description was so vague that it could have been anyone. You might generate reasonable doubt with such an approach, but a jury might also believe that the totality of the evidence suggests your client is guilty, even if each piece can be explained away individually.

The second approach is to argue that the suspect could not possibly have committed the crime: He has an iron-clad alibi for the time of the crime, or he was physically incapable of committing the crime because a childhood injury left him with only one functioning arm. These are examples of "in-principle" arguments. They imply that the accumulated evidence is irrelevant because your client simply cannot have committed this crime. If you can pull it off, this approach is more powerful.

As we shall see in Section 2.2, scientists point to many distinct lines of evidence in making the case for evolution, drawing on distinct sets of facts from every branch of the life sciences. The strength of the case comes not so much from any one line of evidence but from the many concordant lines drawn from numerous branches of science. Anti-evolutionists often dutifully pursue our attorney's first approach of challenging each line individually. They say a great many things in this regard, but scientists invariably find these arguments to be based on faulty facts or logic.

For anti-evolutionists, it would be so much more satisfying to have an in-principle argument against evolution, and mathematics seems like the place to turn to find such a thing. If you can carry out a calculation to show that evolution posits something impossible, or cite an abstract principle that says that large-scale evolution cannot occur, then you are freed from the burden of having to address the various lines of evidence individually. If the numbers do not add up, then the theory is wrong, and that is all there is to it.

Modern anti-evolutionists agree with Reverend Williams in saying that evolution fails the acid test of mathematics, and they have numerous arguments to offer on behalf of that view. We shall spend the remainder of this book considering those arguments.

We shall be forced to conclude that these arguments are entirely inadequate. Modern anti-evolutionists typically avoid the really gross errors of someone like Williams, but that is where the good news ends.

1.6   NOTES AND FURTHER READING

I have provided a detailed summary of the reliance on mathematical arguments in the recent ID literature in a previous paper (Rosenhouse 2016).

For discussions of the nuances and differences among various schools of anti-evolutionist thought, I recommend the books by Numbers (2007) and Scott (2009). For a discussion of the broader political and educational ambitions of anti-evolutionism, have a look at the books by Forrest and Gross (2003), and Berkman and Plutzer (2010).

I remarked that modern evolutionary theory is marked by ferment over details coupled with confidence in the fundamentals, and also that new ideas seem to get introduced faster than they can be assessed and assimilated. A useful reference for both statements is the book edited by Pigliucci and Müller (2010). This book is essentially the proceedings of a conference held in 2008. The conference participants argued that there had been so many advances since the "modern synthesis" of the 1940s, that the time had come to speak seriously of an "extended synthesis." In their preface, Pigliucci and Müller write:

> The modifications and additions to the Modern Synthesis presented in this volume are combined under the term Extended Synthesis, not because anyone calls for a radically new theory, but because the current scope and practice of evolutionary biology clearly extend beyond the boundaries of the classical framework.
>
> *(Pigliucci and Müller 2010, viii)*

The book's contributors were mostly addressing very technical issues of interest to professional biologists, but did not at all challenge the aspects of evolutionary theory relevant to our concerns here.

The basic commitments of YEC are readily available on numerous websites. When I first started, I found the book by Morris and Parker (1987), who promote YEC, to be helpful, both for its clear presentation of creationist thought and for the generally childish tone of its writing. Though this book was written quite some time ago, YEC has not changed importantly in the ensuing years. Foundational texts for presenting the ID perspective are the books by Johnson (1991), Behe (1996), and Dembski (1999). More recent writings will be considered at the appropriate places in this book. Matzke (2009) is an excellent article establishing the fundamental continuity between YEC and ID.

In Section 1.3, I referred to the time I spent attending anti-evolutionist conferences and gatherings. I described my experiences and discussed some of the scientific and theological issues that naturally arose in a previous book (Rosenhouse 2011).

The history of the Scopes trial has been the subject of extensive scholarship. The recently updated book by Larson (2020) has been well received, though personally I find the older account by de Camp (1968) to be more engaging.

The notion of "evolutionary convergence" has been widely debated among scientists. Paleontologist Stephen Jay Gould (1980) famously argued that the long-term trajectory of evolution is influenced by so many variables that were we to "replay the tape of life" starting from its ancient beginnings, it is unlikely that anything like humanity would evolve a second time. Simon Conway Morris (2004), also a paleontologist, demurred, arguing that the prevalence of convergence suggests the evolutionary process is so constrained that a second play of the tape would almost certainly bring us to essentially the same ending point. In my view, Gould had the better argument, but Conway Morris certainly has his points to make. A more recent treatment of these questions can be found in Losos (2018).

2 Evolution Basics

We have a few pages of table-setting ahead of us before we can get down to our main business. For example, we need to discuss what the theory of evolution actually says and why scientists place so much confidence in it.

According to evolutionary theory, modern species arose through a process of descent with modification from earlier species, which themselves arose from even earlier species. By tracing this process backward through time, we eventually arrive at one, or perhaps a small number of, original species that started the process.

The modification occurs because offspring are never perfect copies of their parents. They typically resemble their parents because they have inherited the parents' genes, but they are not exact copies because genetic replication is imperfect. Each new generation therefore exhibits novel genetic sequences and combinations unknown to its forebears.

Some of these novel genes arise from various kinds of mutation, meaning that the sequence of genetic letters in the offspring is different from the corresponding sequence in the parent. In other cases the novelties arise through recombinations of genetic material found in the parents. Since the precise nature and origins of the variations will not be relevant for our purposes, we will, for convenience, refer to these novelties collectively as "mutations." Over long stretches of time, these mutations accumulate to the point where the current generation scarcely resembles its ancient progenitors.

If these ideas are correct, then it follows that any two species alive today share a common ancestor in the perhaps distant past.

Just as you and your siblings share a recent common ancestor (your parents), and you and your first cousins share a more distant common ancestor (your grandparents), so too do modern species share common ancestors. For example, humans and apes presumably share a relatively recent ancestor, humans and horses a more distant ancestor, and humans and lobsters a more distant ancestor still.

This notion sometimes causes confusion since we are in danger of picturing these common ancestors like something from a low-budget horror movie. If you hear that humans and lobsters share an ancestor, and then picture a half-human, half-lobster chimera, then you have the wrong idea. Were the human/lobster ancestor alive for us to examine, it would be unrecognizable as either a human or a lobster. It would instead be a relatively simple and ancient creature, probably resembling something like a worm. That creature would be as surprised as anyone to learn that his descendants gave rise to both humans and lobsters.

A useful model for the sort of gradual change we envision is to imagine that a newborn baby is photographed once an hour until she is 50 years old. Any two consecutive photographs will be indistinguishable, and even photographs taken days apart will look the same. But they are not *actually* the same, since microscopic physical changes are constantly accruing. Photographs taken years apart are very distinguishable, and the first and last in the sequence will not be recognizable as the same person.

The biological analogue is that parents and offspring are immediately recognizable as the same species, as are parents and grand-offspring. But if we had the complete sequence of photographs connecting the most ancient life forms to their modern descendants, we would find the small mutations present in each new generation eventually adding up to very dramatic changes and easily distinguishable species. That is the claim, at any rate. For the baby in the previous paragraph we discussed changes taking place over years, whereas for species we often have to wait for much longer periods of time; however, the principle is the same.

Let us now pause this discussion to confront a potential problem. If we believe that ancient, simple life eventually evolved into modern, complex life, then we must conclude that evolution can produce functional biological systems like wings, eyes, and lungs. It is one thing to claim that an accumulation of small mutations can lead offspring to diverge from their ancestral stock, but it seems like quite another to claim they can lead to modern organ systems.

The answer to this objection is that our description of evolution to this point is incomplete. You see, some of these mutations represent not just random changes from the ancestral stock but actual improvements. Living things compete with one another for scarce resources, and it sometimes happens that offspring possess mutations that give them an advantage in this competition. Offspring blessed in this way will deposit more copies of their genes in subsequent generations, where they can then serve as platforms for still further mutational improvements. The process through which gene frequencies change by the differential survival and reproduction of their bearers is known as "natural selection."

For example, we might imagine that a favorable mutation in a population of ancient, sightless organisms led to one individual possessing a bit of light-sensitive pigment. This individual and its descendants would then be able to orient themselves with respect to a light source, which in most cases would give them a survival advantage over competitors forced to exist in complete darkness. The genes for light sensitivity would quickly spread through the population. In subsequent generations, such "eyespots" might grow, and then the formation of a small pocket could lead to a rudimentary focusing ability. Any sort of clear fluid could then serve as a primitive lens, and by accumulating such small advantageous changes we are well on our way to evolving modern eyes.

There is substantial evidence to support the gradual evolution of eyes over time, but those details are not relevant to the present discussion. What matters is not the specific details of eye evolution but rather the sort of explanation we are proposing. It is true that a purely random process is unlikely to produce a complex, functional structure like an eye. But we can readily imagine a random process producing organisms that are very slightly better than their competitors at gathering information from ambient light. Natural selection will then ensure that these improvements are preserved in subsequent generations, where they can serve as a basis for further small improvements.

This, then, is the theory of evolution by natural selection, the one that scientists are routinely heckled for studying: All modern life forms arose by a process of descent with modification from ancient forebears. That portion of evolutionary change that causes organisms to become better adapted to their environment is the result of natural selection. In particular, complex, functional structures arise primarily through the accumulation of small advantageous changes, with natural selection preserving the early stages while waiting for the next improvement to arise.

This account is sufficient for our purposes because it covers those aspects of evolutionary theory that are challenged by anti-evolutionists. However, for the sake of completeness we should mention that what I have described is only a small subset of the entirety of modern evolutionary theory. There is *far* more to the subject than just common descent and natural selection. After all, my brief summary of evolution required just a few paragraphs, while introductory textbooks manage to go on for many hundreds of pages.

That acknowledged, we have what we need for the discussions ahead, so let us move on.

2.2   THE EVIDENCE FOR EVOLUTION

Scientists are all but unanimous in believing evolutionary theory to be correct, and they point to numerous lines of evidence in support of their view.

The evidence for evolution has been presented at length by numerous writers better qualified than me to do so. I will suggest some useful references in Section 2.7. However, in the interests of making this book as self-contained as possible, I would like to at least highlight some of the main lines of evidence.

First, we have something like our sequence of baby photographs in the form of the fossil record. It sometimes happens that the remains of dead organisms are preserved over vast stretches of time, thereby making them available for study in the present. Geologists have a variety of reliable methods for dating fossils, and that allows us to line them up in chronological order. Each fossil is like a snapshot revealing that an animal with particular physical attributes existed at a certain time and place in natural history. In this way, they reveal something of the sequence of life forms that has played out in natural history, and that sequence is in perfect accord with evolutionary expectations.

In particular, the sequence shows that the earliest life forms were also the simplest, precisely as evolutionary theory leads us to expect. Complexity arose only very gradually over long stretches of time. The history of animals begins with relatively simple sea creatures, which eventually gave way to fish, and only then to amphibians, reptiles, birds, and mammals. Whenever sufficient fossils exist to trace a lineage backward through substantial periods of time, we find, without exception, that modern forms arose as the end result of a coherent historical sequence of predecessors.

Moreover, there are many instances of transitional forms that bridge the gaps between major groups of organisms. There are early amphibian fossils that look very much like fish and later amphibian fossils that look very much like reptiles. So copious are the fossils connecting reptiles to mammals that it is commonplace to speak of mammal-like reptiles (though we should note that for technical reasons paleontologists prefer not to use this term). Evolutionary theory predicts that modern whales are the descendants of land-dwelling mammals, and sure enough there is a detailed sequence of intermediates to show precisely how that happened. Modern humans are preceded by a long sequence of predecessors that become distinctly more ape-like as we go backward in time. These are just a few examples. There are many others.

Absent evolution, there is no reason to expect any coherent pattern at all to the fossil record. We can imagine hypothetical fossil records in which the major forms of life appeared simultaneously with no subsequent change, or in which humans appeared first ahead of fish and reptiles. We can imagine the most complex life forms appearing right at the start, or a random assemblage of forms showing no discernible patterns whatsoever. The actual fossil record is nothing like this. It exhibits instead a sequence entirely consistent with evolution, and that is surely a major victory for the theory.

Moving on, we could also look at the anatomy of modern animals for evidence of evolution. Just as a seemingly healthy adult might bear scars that speak of past infirmities, so too might animals bear vestiges of their evolutionary history.

Indeed, such vestiges are ubiquitous. We have mentioned that whales descended from land-dwelling mammals, and, indeed, they have rudimentary pelvic bones that speak of this history. Since pelvic bones attach legs to bodies, why do whales have them? Humans have an appendix whose primary modern function seems to be to get infected, burst, and kill anyone not fortunate enough to have the necessary surgery. Surely some clue to its origin is gleaned from its strong resemblance to a shriveled up version of an organ other mammals use to digest grass and leaves. For that matter, humans have tailbones, so named for the structures they supported in our evolutionary ancestors. You could literally pick any modern organism and find similar vestiges of an evolutionary past.

These vestiges extend to the molecular and genetic levels. Humans and apes are unable to synthesize vitamin C, but evolution holds that we are the descendants of animals that possessed this function (just as most modern animals other than humans and apes possess it). Interestingly, our genomes possess a broken version of the necessary gene for making vitamin C. If we are not the descendants of creatures with a functional version of that gene, then what is it doing there? Animals are replete with these "pseudogenes," meaning broken genes that are very similar in DNA sequence and chromosomal location to functional genes in related animals. This makes sense under a hypothesis of evolution, but it is hard to explain as the product of intelligent design.

To see why, let us use an analogy from human engineering. Automobiles used to employ a carburetor for combining fuel and air in the proper ratio to achieve combustion within an engine. Carburetors were subsequently replaced with a more efficient technology known as a fuel injection system, and for that reason carburetors are no longer found in modern automobiles. It would be silly for an engineer to build a modern car with a useless, nonfunctional carburetor wasting space underneath the hood, but that is basically what pseudogenes are. They are very common in the genomes of organisms.

Sticking with genetic evidence for the moment, we can also look at so-called retroviral scars, by which we mean remnants of past infections that are sometimes found in modern genomes. Humans and apes share many such scars, each appearing at precisely the same place in their genomes. It is possible that this is sheer coincidence, but it is more likely that we all inherited this scar from the common ancestor of humans and apes. For an analogy, suppose that each of ten siblings possessed a rare genetic anomaly. Would you assume they all independently got the same random mutation, or would you assume they all inherited it from their parents?

Let us shift now to comparative anatomy. As we survey the animal kingdom, it is commonplace to find similar structures performing very different functions. Darwin himself expressed the point like this in *On the Origin of Species*:

> What can be more curious than that the hand of a man, formed for grasping, that of a mole for digging, the leg of the horse, the paddle of the porpoise, and the wing of the bat should all be constructed on the same pattern, and should include the same bones, in the same relative positions?   *(Darwin 1859, 415)*

His point was that this similarity is precisely what you expect from an evolutionary process that modifies existing structures for new purposes. The human, mole, horse, porpoise, and bat inherited the basic arrangement of bones from their common ancestor. This arrangement was then modified by evolution for the distinct modes of life in which these lineages found themselves.

Darwin focused on the forelimbs of various mammals, but he could as easily have taken the skeleton as a whole. Our skulls, for example, show the same general arrangement of bones across all mammals. Biologist Richard Dawkins writes:

> [Y]our ... skull contains twenty-eight bones, mostly joined together in rigid 'sutures', but with one major moving bone (the lower jaw). And the wonderful thing is that, give or take the odd bone here and there, the same set of twenty-eight bones, which can clearly be labeled with the same names, is found across all the mammals. ... [T]he pattern of resemblances among the skeletons of modern animals is exactly the pattern we should expect if they are all descended from a common ancestor, some of them more recently than others. *(Dawkins 2009, 294–295)*

Further evidence comes from embryology. I alluded earlier to the human tailbone, and, indeed, early human embryos possess tails. (It even occasionally happens that a baby is born with a tail). We form structures like gills in the early stages of development, only to have them later change into structures related to breathing air. In this and in many other ways, the obvious inference is that our program of embryological development arose by modifying and adding to fish and reptile programs. We could have chosen the embryological program of any other modern animal to make the same point.

We are only just getting started. I have presented a few simple examples to illustrate certain lines of evidence, but in each case it would have been simple to produce dozens more. Further evidence can be drawn from fields such as biogeography, genetics, and molecular biology; I have omitted such examples entirely due to lack of space. You can study any aspect of any animal from the perspective of any branch of the life sciences, and evidence of evolution will all but smack you in the face. And when you really undertake to study biology in a serious way, when you really immerse yourself in the minutiae of what biologists have discovered about life, you quickly become impatient with those who deny or distort this evidence.

In particular, you come to understand why scientists write at book length presenting the evidence for evolution, and still come away feeling they have done a disservice to the case that could have been made were there no concerns for page count or for a reader's patience. You understand why the prominent twentieth-century biologist Theodosius Dobzhansky (1973) could title an article, "Nothing in biology makes sense except in the light of evolution," and have the scientific community nod vigorously and in unison in response. And you come to understand why professionals regard evolution not as some eccentric theory or wild hypothesis but simply as an obvious fact about natural history. In short, you understand why evolution is taken as an entirely uncontroversial starting point for subsequent research.

Why, then, the relentless heckling from the sidelines?

Much of the opposition to evolutionary theory is born from religious concerns. Whereas religion commonly teaches that we are the intentional creation of a God who loves us, evolution suggests we are just an accidental byproduct of a lengthy and gruesome evolutionary process. It is possible that these are different aspects of the same reality, but it is not obviously so.

However, not *all* of the heckling can be laid at religion's doorstep. Evolution really does seem to run afoul of two basic precepts of common sense. We will devote Sections 2.3 and 2.4 to dispelling them.

2.3 THE UNBRIDGEABLE GAPS ARGUMENT

The first is the plain and simple fact that in nature, like begets like. Dogs give birth to dogs and cats give birth to cats. Period. Some variation around the mean is acceptable, but suggesting that one sort of life can become another sort of life seems unlikely. Life is characterized by unbridgeable gaps between kinds of life, and any theory that claims otherwise, the argument runs, can be dismissed on that basis alone. We shall call this the "unbridgeable gaps" argument.

The notion of sharp boundaries between types of life, however, breaks down upon inspection. Even sticking with dogs and cats, we find that cheetahs are cats with some dog-like features, and hyenas are very dog-like, but also possess cat-like features. One hardly needs Darwin to point out the similarities between humans and apes, and then the similarities of both to monkeys. Reptiles and mammals seem entirely distinct, yet there is the platypus to suggest otherwise. Air-breathing animals are said to have evolved from fish, and sure enough we find several species of "lungfishes" – fish with lungs – to show us how the intermediates might have appeared.

Moreover, biologists have provided evidence for many claimed cases of "ring species." By this we mean a connected sequence of neighboring populations, each of which interbreeds with the next in line, but in which the end populations are no longer capable of interbreeding despite living in the same area. The concept of a "ring species" is controversial among biologists, and some of the claimed examples have been challenged in the literature. However, the empirical evidence strongly supports my point regardless of whether or not a specific, claimed example perfectly satisfies the definition of a ring species. Specifically, numerous small variations that are barely noticeable between adjacent links in a chain quickly add up to something significant. In light of such examples, species boundaries do not seem so inviolable after all.

Thus, even arguing solely from the visible attributes of modern organisms, we see more evidence for continuity than we do for perpetually distinct forms. This impression extends to the molecular level, where we find that all organisms use the same genetic code, with only a very small number of minor exceptions, which themselves occur in a pattern consistent with common descent. (That is, the variations in the code are not found scattered randomly throughout the animal kingdom, but instead occur in patterns that suggest that some grouping of animals inherited the change from a single common ancestor.) Moreover, the same genes seem to be implicated in the same organs across the different forms of life. Increasingly, scientists are able to pinpoint the precise genetic changes underlying major transitions in evolution.

We can readily imagine an alternate reality in which we find numerous genetic codes, corresponding to some plausible notion of different kinds of organisms. It might have turned out that dogs had one code, cats another, and birds a third, for example, and this would have been strong evidence for unbridgeable gaps and against evolution. That is not what we actually find, however.

The unbridgeable gaps argument takes another blow from the prevalence of transitional forms in the fossil record. Here we find, for example, dinosaurs with feathers, fish with feet, and snakes with legs. Frequently, fossils are available to show us the actual bridges between seemingly unbridgeable gaps.

Therefore, a minimal response to the unbridgeable gaps argument is that it is unpersuasive, while a maximal response is that people who put it forth do not know what they are talking about.

2.4   THE COMPLEX STRUCTURES ARGUMENT

The second objection is that natural forces lead only to weathering, erosion, and decay. They do not build complex, functional structures. Our everyday experience is that structures break down unless energy and intelligence are expended to maintain them. How then can we argue that evolution, alone among natural processes, has the power to fight this tendency? How can it produce order where there was only chaos before?

Darwin took this argument seriously, writing:

> To suppose that the eye with all its inimitable contrivances
> for adjusting the focus to different distances, for admitting

different amounts of light, and for the correction of spherical and
chromatic aberration, could have been formed by natural
selection, seems, I freely confess, absurd in the highest degree.

(Darwin 1859, 217)

The American botanist Asa Gray, who was largely supportive of
Darwin's ideas, likewise believed the theory foundered on this point.
In a lengthy review of *The Origin of Species* published shortly after
the book's publication, Gray wrote:

Accordingly, as he [Darwin] puts his theory, he is bound to
account for the origination of new organs, and for their diversity
in each great type, for their specialization, and every adaptation of
organ to function of structure to condition, through natural
agencies. ... Here purely natural explanations fail.

(Gray 1876, 52)

Let us call this the "complex structures" argument. We shall
devote this section to some replies.

Anti-evolutionists often argue that we immediately recognize
intelligent design when we see functional machines. When we see a
pocket watch or a telescope, we do not for a second believe they were
created by any purely natural process. They go on to argue by analogy
that functional systems in organisms must also be the products of
design. Biologists reply, with considerable justice, that this argument
is based on a bad analogy between human contrivances and biological
adaptations.

We can turn the logic around on the anti-evolutionists by noting
that some systems are immediately recognized as having resulted
from a lengthy, historical process. When a system is designed with
weird kludges and bizarre arrangements of parts, we know the system
did not arise in one step from the mind of an engineer.

For example, the house I live in is close to 70 years old and
features a byzantine arrangement of water pipes that go snaking
around the basement. Some of these pipes are part of the heating

system; boiling water gets sent through the pipes, which then radiate heat to the rest of the house. Others are part of the plumbing system that services the kitchen and bathrooms. If you care to visit me and examine the pipes, you will quickly notice that their arrangement seems a little daft. There is one collection of pipes *over here,* and then what looks like a separate arrangement *over there,* and you will quickly decide that the engineer who came up with this must have been a little daft himself.

Of course, the explanation is obvious. One of the house's previous owners decided to build an extension onto the original structure. Extending the heating system into this new space required additional pipes, which were then clumsily attached to the existing system. A different previous owner decided to install a second bathroom, and the system of pipes needed to be extended once again. In each case, additions had to be added to an already existing structure since it was completely impractical to start over. It all works well enough, but an engineer starting from scratch could easily devise a more efficient system.

Complexity by itself is a poor argument for intelligent design since such a result can also arise through a series of small steps when an old system is awkwardly modified to serve new ends. Weird, nonsensical complexity is a strong argument for evolution, not design.

I was introduced to this principle many years ago, when I was ten. I was driving with my father to his office. The drive took us down a narrow, two-lane road that passed a series of farms. I liked this part of the drive because one of the farmers owned a pair of llamas, and they were usually out grazing as we drove by. My father always slowed down at this point so I could get a good look.

Eventually our road came to a T at a major highway. We wanted to go straight through, but this was impossible because the continuation of our road was not directly across from us. Instead, it was offset by a small distance. Starting from a full stop, we had to first make a hard right turn onto the highway and then almost immediately make a hard left off of it. This was a dangerous maneuver, since the highway

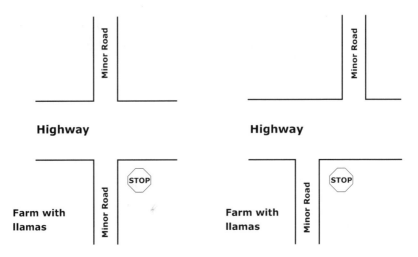

FIGURE 2.1 (left) A sensible arrangement of roads of the sort a civil engineer would devise. (Right) An absurd and potentially dangerous arrangement that only makes sense when you understand the historical events leading up to it.

always had lots of fast-moving traffic, and because a slight bend in the road made it difficult to see what was coming. Unsurprisingly, crashes were common at that intersection. The arrangement is shown in Figure 2.1.

I asked my father why the roads were designed that way. Even at ten I understood that it made more sense for the roads to be laid out like a plus sign, so that we could go straight through. My father smiled and explained that no one had designed this arrangement of roads. What actually happened was this: The major highway had been built at a time when the surrounding land was largely undeveloped. Gradually, separate towns arose on either side of it. As the towns grew, each decided independently that it needed a connection to the highway. So each built its own connection, and that is why the roads did not line up.

These examples perfectly represent the essential distinction between intelligent design and evolution. Viewed as the product of intelligent design, the arrangement of pipes in my house or the arrangement of roads on that drive with my father seemed ridiculous. But viewed as the outcome of a long historical process, everything made sense. Paleontologist Stephen Jay Gould coined the phrase "the senseless signs of history" to make the same point. Elaborating, he wrote:

> [I]deal design is a lousy argument for evolution, for it mimics the postulated action of an omnipotent creator. Odd arrangements and funny solutions are the proof of evolution – paths that a sensible God would never tread but that a natural process, constrained by history, follows perforce. *(Gould 1980, 20–21)*

Anthropologist Alan Rogers explains the point well. Using the analogy of a gardener whose water hose has become hooked around a tree, he writes:

> As our ancestors evolved from fish into amphibians, reptiles, and mammals, their bodies changed enormously in size and shape. This required corresponding changes in the various tubes and wires – arteries, veins, nerves, and so on – that run throughout our bodies. In many cases these tubes became stretched around some obstacle, confronting selection with a dilemma like that of the gardener .... All too often, it failed to do the sensible thing. Rather than walking back around the tree, selection got another length of hose. *(Rogers 2011, 54)*

Rogers illustrates this with two examples. The first is the vas deferens in human males, which carries sperm from the testis to the penis. This tube is far longer than it needs to be because it is hooked over the ureter, which connects the kidneys to the bladder. As Rogers notes, "It works, but it is not something an engineer would be proud of." (Rogers 2011, 56)

His second example is the recurrent laryngeal nerve. His description is worth quoting at length. Referring to a certain arrangement of nerves and arteries in fish, he writes:

In the nerves of a fish, this leads to a simple wiring diagram. In ours, it leads to a tangle that generations of anatomy students have learned to dread. ... [The recurrent laryngeal nerve] starts in the head, travels down to the chest where it loops around an artery, and then travels back up to the throat. In humans, it ends up only a few inches from where it started. ... In fish, the nerve and artery both feed the rearmost gill arch. Over time, the artery descended into the chest, and the nerve went with it. In giraffes, the nerve is 20 feet long, yet the direct route is only a foot. Engineers get fired for this sort of thing.   *(Rogers 2011, 57)*

We should emphasize that the issue in both of these cases, and in the dozens more that could quickly be produced to illustrate the same idea, is not so much bad design as it is weird design. Engineers often have to make trade-offs and compromises when solving practical problems. In such situations, something that initially appears to be poor design can instead be understood as the best possible solution given other constraints.

That is not what we are talking about here, as the example of Figure 2.1 illustrates. The weird arrangement of roads did not arise because a civil engineer faced design constraints. Rather, the arrangement is such that no engineer would even have considered it. The roads are incomprehensible until you consider the historical process leading to them. Likewise for Rogers' two examples. The absurdity of the giraffe's laryngeal nerve or the human vas deferens are in no way mandated by other design constraints, and they are nothing even the world's worst engineer would ever have dreamed up. Viewed as products of engineering they are ludicrous, but they fall right into place as soon you understand the evolutionary process that led to them. After a discussion of the giraffe's laryngeal nerve, biologist Richard Dawkins aptly summarizes the situation:

[I] soon realized that, where imperfection is concerned, the recurrent laryngeal is just the tip of the iceberg. The fact that it takes such a long detour drives the point home with particular

force. ... But the overwhelming impression you get from
scavenging any part of the innards of a large animal is that it is a
mess! Not only would a designer never have made a mistake like
that nervous detour; a decent designer would never have
perpetrated *anything* of the shambles that is the criss-crossing
maze of arteries, veins, nerves, intestines, wads of fat and muscle,
mesenteries and more. To quote the American biologist Colin
Pittendrigh, the whole thing is nothing but a 'patchwork of
makeshifts pieced together, as it were, from what was available
when opportunity knocked, and accepted in the hindsight, not the
foresight, of natural selection.'   *(Dawkins 2009, 371)*

Living things possess many complex adaptations. Without
exception, they show clear signs of having resulted from a lengthy,
stepwise, historical process. They show the senseless signs of history.
They are never pristine creations from nothing, as we might expect if
they were designed by a powerful intelligence, but instead are invariably cobbled together from parts that are just lying around, so to speak.

The concept is well illustrated by the various contrivances used
by orchids to attract insects. Darwin made a comprehensive study
of such structures, and he made the following observation (note that
"homology" refers to the use of the same parts for different purposes
in different species):

If, indeed, [the reader] should care to see how much light, though
far from perfect, homology throws on a subject, this will, perhaps,
be nearly as good an instance as could be given. He will see how
curiously a flower may be moulded out of many separate organs, –
how perfect the cohesion of primordially distinct parts may
become, – how organs may be used for purposes widely different
from their proper function, – how other organs may be entirely
suppressed, or leave mere useless emblems of their former
existence. Finally, he will see how enormous has been the total
amount of change the simple parental or typical structure which
these flowers have undergone.   *(Darwin 1862, 289)*

Any other complex adaptation could be used to make the same point. For example, the flagellum used for propulsion by certain types of bacteria is composed of numerous proteins working together to carry out a definite purpose. In considering how such a structure might have evolved gradually, we are aided by the fact that nearly all of these proteins have other functions throughout the cell. Biochemist Ian Musgrave writes:

> [W]e now know that between 80 and 88 percent of the eubacterial flagellar proteins have homologs with other systems, including the sigma factors and the flagellins. Homologies between a few of the rod proteins and nonflagellar proteins have not been found yet, but they appear to be copies of each other and related to the hook protein. In the end, there is not much unique left in the flagellum.   (Musgrave 2004, 81)

(We will have much more to say about the evolution of the flagellum in Section 5.9.)

In other words, in every case we find that the complex adaptation has precisely the structure it would need to have for natural selection to be a viable hypothesis. Selection can only craft adaptations out of raw materials that are already there, and all of the myriad adaptations studied to date have turned out to be so crafted.

This is a good start, but we might still wonder about the intermediate stages linking primordial simplicity with later complexity. For this we can often appeal to the fossil record, which in some instances is sufficiently detailed to permit strong conclusions to be drawn. An example is the evolution of flight in birds. Paleontologist Alan Gishlick writes,

> The fossil record of the evolution of avian flight is extensive and constantly growing; in particular, we now have a detailed record showing how the skeletal and muscular systems were modified along the route to flight. What we see in this record is that all of

the skeletal, ligamentous, and muscular features just discussed
arose gradually along the lineage leading to birds.

(Gishlick 2004, 66)

In some cases, fossil evidence can be backed up with findings
from genetics or embryology. An example is the hearing apparatus in
mammals, which consists of three small bones that conduct sound
from the ear drum to the inner ear. It is a complex, multi-part,
functional structure. However, long before Darwin, anatomists and
embryologists had noticed that our inner ear bones were homologous
to similar bones found in the reptilian jaw joint. Subsequent fossil
finds showed how it was possible for a jaw bone to be coopted into
an ear bone. Specifically, reptile fossils were found with a double jaw
joint, and this redundancy made it functionally possible for one of the
bones to be coopted. That this is not just a theoretical possibility, but
actually happened, is then shown by the formation of these bones in
embryological development. Stephen Jay Gould writes:

> Thus, every mammal records in its own embryonic growth the
> developmental pathway that led from jawbones to ear bones in its
> evolutionary history. In placental mammals, the process is
> complete at birth, but marsupials play history postnatally, for a
> tiny kangaroo or opossum enters its mother's pouch with future
> ear bones still attached to, and articulating, the jaws. The bones
> detach, move into the ear, and the new jaw joint forms – all during
> early life within the maternal pouch.   *(Gould 1993, 105)*

Comparative anatomy in the present can also provide powerful
clues about the intermediate stages of a complex structure. Earlier
I suggested the possibility that eyes evolved through various stages
starting with a mere spot of light-sensitive pigment. The plausibility
of such a scenario is increased when we note that eyes in every
phase of transition exist in the present, thus proving that they are
fully functional. Genetics and molecular biology can then establish

homologies among the proteins involved in vision, with the result being a convincing account of how the eye evolved.

Darwin expressed the basic logic very well. Earlier I quoted him expressing incredulity at the thought that the eye could have evolved gradually. Continuing from that quotation, Darwin writes:

> Yet reason tells me, that if numerous gradations from a perfect
> and complex eye to one very imperfect and simple, each grade
> being useful to its possessor, can be shown to exist; if further, the
> eye does vary ever so slightly, and the variations be inherited,
> which is certainly the case; and if any variation or modification in
> the organ be ever useful to an animal under changing conditions of
> life, then the difficulty of believing that a perfect and complex eye
> could be formed by natural selection, though insuperable by our
> imagination, can hardly be considered real.   *(Darwin 1859, 155)*

A century and a half of research has put meat on the bones of Darwin's plausibility argument, to the point where today the evolution of eyes is not considered to be mysterious.

To these theoretical considerations we can add two sorts of experimental evidence. First, field studies of animals in the wild consistently show that natural selection is a powerful force. It can effect considerable changes in modern organisms over very short time spans. Second, animal and plant breeders have been enormously successful in modifying species by sieving naturally occurring variations in whatever direction they desire. It is in this manner that animal breeders, for example, have produced the riot of modern dog breeds starting from an ancestral wolf. If a Great Dane, a Dachshund, and a Chihuahua were known only from fossils, they would probably not be placed in the same genus, much less the same species. Yet all arose by a variation/selection mechanism in a relatively short amount of time.

These two lines of evidence amount to a powerful proof of concept for the idea that naturally occurring variations, when sieved through a consistent selection mechanism, can effect major change in modern organisms over short periods of time. In light of this evidence, the burden is on the other side to show there is a fundamental barrier to the quantity of change possible over vastly longer periods of time.

Assembling the pieces, the refutation of the "complex structures" argument is this: A great many complex structures have been studied, and in every case they have just the structure they would need to have for natural selection to be a viable hypothesis. Specifically, they are always seen to be cobbled together from readily available parts, and do not appear as *de novo* creations disconnected from anything that came before. Like the system of roads leading to my father's office, they show the senseless signs of history that are indicative of evolution, not design. Moreover, for many specific adaptations, extensive evidence from paleontology, genetics, embryology, and anatomy converge to paint a well-supported account of how they evolved. We also know experimentally that a variation/selection mechanism can result in major changes to modern organisms in short periods of time.

Scientists therefore have a strong basis for concluding that natural selection can craft complex structures.

2.5   IRREDUCIBLE COMPLEXITY

Proponents of ID frequently appeal to a variation on the complex structures argument. They argue that it is not complexity per se that challenges evolution, but instead a particular kind of complexity, which they call "irreducible complexity." Since this concept will play a crucial role in our subsequent discussions, especially in Section 5.7, we will devote this section to it.

In *On the Origin of Species*, Charles Darwin wrote:

> If it could be demonstrated that any complex organ existed which could not possibly have been formed by numerous, successive slight modifications, my theory would absolutely break down. But I can find out no such case.   *(Darwin 1859, 219)*

The concept of irreducible complexity is meant to respond to this challenge. The basic idea is that if a system requires all of its parts to function properly, then it could not have formed by "numerous, successive slight modifications," because the precursor systems, lacking some of their parts, would not be functional.

In various forms, this idea is as old as anti-evolutionism itself. The modern revival of this argument in ID discourse, as well as its application specifically to biochemical systems, is due to biochemist Michael Behe. In his 1996 book, *Darwin's Black Box*, he argued that he had found the systems Darwin overlooked:

> By irreducibly complex I mean a single system composed of several well-matched, interacting parts that contribute to the basic function, wherein the removal of any one of the parts causes the system to effectively cease functioning. An irreducibly complex system cannot be produced directly (that is, by continuously improving the initial function, which continues to work by the same mechanism) by slight, successive modifications of a precursor system, because any precursor to an irreducibly complex system that is missing a part is by definition nonfunctional. An irreducibly complex biological system, if there is such a thing, would be a powerful challenge to Darwinian evolution.   *(Behe 1996, 39)*

Most of his book is then given over to describing various biological systems that have this property of interlocking complexity, such as the human blood clotting cascade or immune system. In each case, several distinguishable parts work together to carry out some function, and if any one part is removed or damaged the system ceases to work.

Behe is not humble regarding the conclusion of his argument:

> The result of these cumulative efforts to investigate the cell – to investigate life at the molecular level – is a loud, clear, piercing cry of "design!" The result is so unambiguous and so significant

that it must be ranked as one of the greatest achievements in the history of science. The discovery rivals those of Newton and Einstein, Lavoisier and Schrödinger, Pasteur and Darwin.

(Behe 1996, 233)

To better understand Behe's argument, recall that evolutionists rely on the notion of stepping-stones to explain complex adaptations, where each stepping-stone represents a functional structure that could be preserved by natural selection. For example, we briefly mentioned a few of the likely stepping-stones involved in evolving an eye – starting with a light-sensitive spot, moving on to a spot with a small pocket, then an eye with a primitive lens, and so on – but each of those steps represented a functional structure. This is critical because natural selection only understands immediate reproductive success. It will not preserve a worthless structure in the hope that later mutations will transform it into something useful.

Behe is arguing that when a system is irreducibly complex there are no stepping-stones. This is because at least some of those stones would have to be structures that were missing a part relative to the modern system, and they would therefore be nonfunctional. Natural selection would want nothing to do with them. He argues that since such a system could not have evolved gradually through functional intermediates, and since it could not have arisen fully formed just by chance, the only remaining option is intelligent design.

Behe's writing is cocky and full of swagger, but even before discussing the details we can be very skeptical of his argument. It is clever marketing to refer to these systems as "irreducibly complex." A more accurate description is "easily broken," and his argument could then be rephrased like this: "The prevalence of easily broken systems in nature is strong evidence of intelligent design." In *that* form the argument is not terribly persuasive.

We have noted that much of the anti-evolutionary discourse is based on analogies to human engineering. They argue that just as humans can build machines that we immediately recognize as having arisen from intelligence and not from natural causes, so too can we be certain that nature's machines were designed by an engineer of vastly greater intelligence. However, irreducible complexity is more accurately seen as another instance of the senseless signs of history. Human engineers would not build a complex system in such a way that it fails catastrophically if the slightest thing goes wrong. If biology presented us with resilient systems complete with backups, fail-safes, and redundancies, then we might more naturally suspect intelligent design by a master engineer. In contrast, systems balanced on a knife-edge of functionality suggest something closer to a failing grade in a first-year engineering course.

In the human context this is not a trivial concern. The fragility of some of Behe's favorite examples, such as the human blood clotting cascade or the immune system, leads to tremendous human misery and misfortune. People who suffer from hemophilia have low levels of one or another of the factors needed to clot blood at the site of a wound, and because of this they live in terror of trivial injuries other people would shrug off. They might have appreciated a more resilient clotting system, one that does not fail when one part is a little sub par. Likewise, there are over eighty autoimmune diseases, in which the immune system for some reason attacks its own body, and I invite you to lecture the numerous victims of these diseases about the brilliance of the engineer responsible for it.

It would seem, then, that if Behe's argument has any force at all, it can only be because of his central claim – that an irreducibly complex system cannot arise in a stepwise manner through gradual evolution. If this claim is correct, then we might feel forced to the conclusion of intelligent design, the considerations of the last few paragraphs notwithstanding.

However, the claim is plainly not true. A moment's thought is sufficient to come up with scenarios through which an irreducibly complex system could arise gradually.

For example, an interdependence of parts can arise through the removal of redundancy. A standard example is a stone arch. The arch cannot support itself until the capstone is placed on top, and the capstone has nothing to rest on until the arch is complete. During construction a scaffolding supports the structure. When the scaffolding is removed, the result is an interdependence of the remaining parts. Applying Behe's logic, we should look at the finished structure and conclude that it just appeared from nothing, fully formed, with no intermediate stages.

Another possibility is that changes that are merely improvements at first can later become essential because the environment changes around them. Telephones were a luxury item when they were first introduced, but today they are considered so vital that most of us carry one with us everywhere we go. If the world's telephones suddenly disappeared from the face of the earth, civilization as we know it would be very seriously compromised. The biological analog is that a mutation might initially confer some small advantage on its bearer, but later become essential as the environment changes, and other genes mutate, around it.

It is also possible for numerous systems to evolve in tandem. A modern city can be viewed as irreducibly complex, with separate systems for transportation, power generation, waste removal, communications, banking, and others besides, and these systems can themselves be viewed as being composed of discrete parts. There are many components which, if they were suddenly knocked out, would quickly cause the city to effectively cease functioning, but there is no mystery to how the modern city arose gradually over long periods of time. Each of those modern systems originally existed in less effective, more rudimentary forms. Improvements in one system then led to improvements in others, with the result being the modern interdependence of parts. There is no reason in principle why an irreducibly complex biological system could not evolve by a similar process.

These possibilities reveal a further problem with Behe's logic. He seems to think that a complex system evolves by the sequential addition of discrete parts. Philosopher Philip Kitcher writes, specifically using the example of the bacterial flagellum, frequently appealed to in ID literature:

> We are beguiled by the simple story line Behe rehearses. He invites us to consider the situation by supposing that the flagellum requires the introduction of some number – 20, say – of proteins that the ancestral bacterium doesn't originally have. So Darwinians have to produce a sequence of 21 organisms, the first having none of the proteins, and each subsequent organism having one more than its predecessor. Darwin is forlorn because however he tries to imagine the possible pathway along which genetic changes successively appeared, he appreciates the plight of numbers 2-20, each of which is clogged with proteins that can't serve any function, proteins that interfere with important cellular processes. These organisms will be targets of selection, and will wither in the struggle for existence. Only number 1, and number 21, in which all the protein constituents come together to form the flagellum, have what it takes. Because of the dreadful plight of the intermediates, natural selection couldn't have brought the bacterium from there to here.
>
> The story is fantasy, and Darwinians should disavow any commitment to it.   (Kitcher 2007, 88)

The story is a fantasy because cooption of function is a commonplace of evolution, as we have noted. In other words, finding the stepping-stones to an irreducibly complex structure does not involve producing one new protein after another, but instead, in large measure, involves repurposing already-existing proteins to new functions.

Now, even if these replies were just so much armchair theorizing, they would still be sufficient to refute Behe's argument. He put forth an in-principle argument: If a system has a certain interdependence of parts then it could not evolve gradually. It is therefore appropriate to offer an in-principle reply: Your claim is incorrect. Here are some ways such a system could arise through evolution.

However, as it happens, the issues I raised in the last few paragraphs were pointed out to Behe in numerous reviews right after the publication of his book. Evolutionary biologists had no difficulty thinking up abstract scenarios to explain irreducible complexity because they had numerous concrete biological examples to point to. We saw some of them in Section 2.4. For example, the evolution of the three bones of the mammalian inner ear involved a loss of redundancy in reptilian precursor systems, and Darwin's orchids show the possibilities of modifying pre-existing structures in tandem and of the cooption of function of pre-existing parts. Biologists can provide plausible scenarios for numerous other such systems as well. The general sorts of mechanisms through which evolution crafts complex structures are well understood, and, as we have discussed, there is substantial circumstantial evidence to show how these mechanisms have played out in practice.

In short, scientists responded to Behe by saying, effectively, "Irreducible complexity in the present tells us nothing at all about functional precursors in the past because there are many ways for an interdependence of parts to arise gradually. Moreover, there are many specific 'irreducibly complex' systems where we have very strong evidence to show us what the stepping-stones actually were."

For evolutionary biologists, there was nothing remotely new in Behe's argument. Darwin himself already recognized the importance of the removal of redundancy in forming complex structures:

> We should be extremely cautious in concluding that an organ could not have been formed by transitional gradations of some kind. Numerous cases could be given amongst the lower animals of the same organ performing at the same time wholly distinct functions ... In such cases natural selection might easily specialise, if any advantage were thus gained, a part or organ, which had performed two functions for one function alone, and

thus wholly change its nature by insensible steps. Two distinct organs sometimes perform simultaneously the same function in the same individual ... In these cases, one of the two organs might with ease be modified and perfected so as to perform all the work by itself, being aided during the process of modification by the other organ; and then this other organ might be modified for some other quite distinct purpose, or be quite obliterated.

(Darwin 1859, 220)

Darwin was hardly the only one to notice. In 1918, biologist H. J. Muller, a future Nobel laureate, wrote:

[T]hus a complicated machine was gradually built up whose effective working was dependent upon the interlocking action of very numerous different elementary parts or factors and *many of the characters and factors which, when new, were originally merely an asset finally became necessary because other necessary characters and factors had subsequently become changed so as to be dependent on the former.* It must result, in consequence, that a dropping out of, or even a slight change in any one of these parts is very likely to disturb fatally the whole machinery; for this reason we should expect very many, if not most, mutations to result in lethal factors ...

(Muller 1918, 463–464, emphasis in original)

Muller's point is that natural selection all but inevitably crafts complex systems showing an interdependence of parts.

Working evolutionary biologists have all of this material at their fingertips, which is why they responded so negatively to Behe's book. What Behe described as an astonishing discovery that shrieks design, biologists recognized as a trivial problem that had been solved close to a century earlier.

"Irreducibly complex" systems not only pose no challenge to evolution, they are actually the expected outcome of prolonged natural selection. The claim that such systems cannot evolve gradually is manifestly false, leaving us only with the fact that irreducible complexity represents appalling design, at least as judged by the standards of human engineers.

Therefore, the prevalence of irreducibly complex structures is strong evidence for evolution, and it is strong evidence against intelligent design.

2.6   PAYING A PRICE FOR BEING WRONG

Anti-evolutionists will be quick to point out what I did *not* do in the previous two sections. I did not point to a specific population of, say, eyeless creatures, and then show you a film of their descendants gradually evolving eyes. Nor did I do anything comparable for any other complex structure.

The evolutionary process is such that complex adaptations take a very long time to develop. Consequently, we have to rely on circumstantial evidence in making our case. Such evidence can be very compelling, and it can accumulate to the point where reasonable doubt is eliminated, but it is circumstantial nevertheless.

The anti-evolutionists are certainly free to point this out, but doing so represents a significant scaling-back of their ambitions. Let me explain what I mean.

In any discussion of the evolution/creation controversy, it is all but inevitable to mention William Paley's 1802 book *Natural Theology* (Paley 2006). Paley's argument was that the interlocking complexity of organisms could only be explained by reference to God and not by any fully naturalistic process. Most of his book's chapters are occupied with detailed descriptions of complex adaptations in nature.

Paley wrote his book well before Darwin arrived on the scene, and the success of modern evolutionary science has made his argument far less persuasive than it used to be. That aside, Paley's discourse is essentially identical to that presented by modern anti-evolutionists. He uses the complexity of biological adaptations to infer that there is an intelligent agent behind it all.

The aspect of Paley's book that is relevant for us is this: At no point does he argue that scientists, in their professional work, should heed what he is saying. He does not tell scientists how to do their jobs, and he does not argue that his conclusion of design in nature constitutes a theory of which scientists need to be cognizant in carrying out their professional work. In effect, he is saying, "Here are some facts about anatomy uncovered by scientists, and here is a theological conclusion I have drawn from those facts." At no point does he say anything like, "You incompetent scientists are doing it wrong! You better bring an assumption of intelligent design into the lab with you, or science will continue down a dramatically wrong path!"

In this he stands in stark contrast to modern anti-evolutionists, who claim that they are advancing fully scientific theories that should replace evolution in the daily work of practicing scientists. This is true of both YEC and ID.

This returns us to my earlier claim, that anti-evolutionists are scaling back their ambitions in basing their case on the lack of direct evidence for the gradual formation of complex structures. Their big claim was supposed to have been that they had a novel and fruitful approach to biology. For working scientists, that is what makes them potentially interesting (keeping in mind that we are, for now, ignoring the cultural milieu in which these conflicts play out). If instead the anti-evolutionists just want to be truculent and contrarian, insisting the evidence is insufficient to convince *them*, then scientists would just reply with, "Whatever. Believe what you want." They are welcome to sit and pout in the corners of the gym while everyone else is up dancing.

But then the scientists would add, "However, we seem to be getting good results with standard evolutionary theory, so we'll just stick with it until you come up with something better."

I opened this chapter with a brief outline of the evidence for evolution. In so doing, I followed tradition by presenting the evidence in the manner of an attorney prosecuting a case. I gestured toward broad classes and general lines of evidence from paleontology, anatomy, molecular biology, and embryology. This is all good stuff, but it has a tendency to seem rather abstract, and it misses something important. Specifically, it misses the point that scientists have a job to do, and this job has nothing to do with promoting worldviews. Rather, their job is to get tangible results on practical problems. Theories are tools they use while doing their job, and sheer practicality forces them to stick with tools that work and to discard the tools that do not work. Evolution remains the dominant paradigm in science because it consistently gets good results in practical situations. ID and YEC offer nothing to rival this success.

On scientific questions, if you want to know who is giving you the straight story, you should listen to the people who pay a price for being wrong.

For example, when energy companies hire geologists to help them find new sources of oil and natural gas, they do not look for people with expertise in young-Earth geology. They do not hire the people who say the earth is only 10,000 years old and that Noah's flood was a real event. This is not because of anti-religious bias, but is instead solely because real geology gets results, and young-earth geology does not.

Or consider the shape of the earth. Most people can go their entire lives believing that the earth is flat and never make a bad practical decision because of it. After all, in the early days of human civilization, people mostly believed precisely that. But suppose you run an international airline and have to plan efficient routes connecting one side of the earth to the other. Or suppose you work for NASA, and you have to launch spacecraft that can accurately hit distant targets in space. In *those* contexts it really matters that you get the right answer about the shape of the earth, and you will find that the people who do that sort of work have little sympathy for flat-earth theory.

I mentioned earlier having attended a fair number of anti-evolutionist conferences. At those events, I heard speaker after speaker rail against evolution. I chatted with many audience members entirely convinced of their own erudition, especially as compared to the benighted scientific community. Even more than the manifest errors in the claims and arguments I heard, I was struck by just how cheap and tawdry most of the criticisms were. Preachers can get away with ranting about the evils of evolution, anti-evolution conference speakers can level baseless and scurrilous charges against scientists, and creationists can congratulate each other for their sagacity in private conversations, because in those venues there is no penalty for being wrong.

I had the same reaction to the many occasions on which I read anti-evolution screeds in politically conservative publications. These pieces were nearly always written by people with no particular credentials, for an audience likewise composed primarily of lay people. Since no one expected these folks to produce a result in the field or the lab, they had the freedom to level any charges, no matter how ill-conceived or fallacious, safe in the knowledge they would suffer no harm for being wrong.

But as soon as the conversation moves to a venue where there *is* a price to be paid, as soon as you are in an environment where people cannot afford to mess around, evolution really is the only game in town.

A striking example is the discovery of the fossil *Tiktaalik roseae* in 2004. *Tiktaalik* is a fish that nonetheless has numerous reptilian features, making it a plausible transitional form linking fish to tetrapods (four-legged, land-dwelling animals). Even more interesting than the fossil itself is the reasoning that led to its discovery. Paleontologists studying life's move from water to land had strong evidence from genetics and anatomy to suggest that this transition occurred roughly 350–400 million years ago. They then looked for locations where they could find exposed rocks of that age, believing that was their best chance at finding a helpful transitional form. Sure enough, they found what they were looking for.

If evolutionary theory is as misguided as the critics say, then these researchers got incredibly lucky. On the other hand, paleontologists who let evolutionary thinking guide their research seem to get lucky a lot.

In place of paleontology, I could have used any other branch of the life sciences. Our previous discussions have suggested how evolution makes sense of the data found by geneticists, anatomists, and embryologists, but there are numerous other examples as well. We alluded to some of these in Section 1.1, but they are worth another look.

Ethologists who study animal behavior routinely use game theory models in their work. The idea is to view animals as competitors in a game for scarce resources and their sometimes eccentric behaviors as strategies in those games. Starting from this premise, mathematical models can be developed to make predictions about animal behavior in different situations. Researchers have had enormous success with this approach, and the logic behind many formerly incomprehensible animal behaviors has been revealed. This is interesting because the mathematical models are based explicitly on the assumption that the behaviors arose via prolonged natural selection. The success of the models is a vindication of this assumption.

Epidemiologists routinely apply evolutionary techniques in their work. Phylogenetic analysis, which is the branch of biology concerned with working out the evolutionary relationships among organisms, has been used to combat outbreaks of diseases and to devise treatments for AIDS and influenza, among other examples. The daily practice of medicine has also been affected for the better, now that we better understand the processes through which microbes evolve resistance to antibiotics.

We can also look beyond the life sciences. A common problem in computer science is to search a large space of possibilities for target points matching narrow specifications. For very large spaces a full census, where you just try every point, is impractical, and random sampling is also unlikely to bring success. An algorithm is needed to guide the search, which in this context can be viewed as a strategy for deciding precisely which points to sample. Since biological evolution can be construed as finding small targets (functional organisms) within a large space of possibilities (all possible combinations of genes), researchers had the idea of mimicking its processes as a search strategy. The result was the field known as "evolutionary computation," which has had many practical successes and remains an active area of research today.

For more than a century and a half, researchers have used evolution to guide their research, and they have been rewarded with one practical success after another. During this same time period, anti-evolutionists just stared off into space and accomplished nothing.

Let us imagine a comparable case. Suppose a well-funded group of activists wages a campaign against hammers. They provide a host of physics-based arguments, complete with equations, claiming to show that hammers do nothing to amplify the force of your arm. They produce slick videos of people injuring themselves with hammers. They warn their followers, none of whom actually work with wood, that the hegemony of hammers is maintained only because tyrannical woodworkers openly despise them and their values.

How might we reply to this? We could certainly take their arguments one at a time. We could point to the errors in the critics' physical arguments, and show that a proper understanding of the relevant equations proves that hammers really do amplify the force from your arm. We could object that the videos only show what happens when you use hammers incorrectly and then instruct people in proper safety precautions. We could expose the smear campaign against woodworkers as just so much scurrilous propaganda. These are all good things, and were this case to play out we would no doubt do all of them.

However, we could also reply by using a hammer to pound a nail into a piece of wood. We could then point to the results and say, with a bemused look, "It sure *seems* like hammers work. Good luck pounding a nail with any other tool."

This book is largely devoted to the first approach. We will dutifully examine the arguments of mathematical anti-evolutionism and show why they do not work. We will take their manifest errors as an opportunity to present some clear thinking about mathematics.

That notwithstanding, the second approach is even more important. Scientific theories are tools that scientists use in their work in
precisely the same way that hammers are tools that woodworkers
use. Theories that work survive, and theories that do not are quietly
discarded. Evolutionary theory works, and that is why scientists stick
with it. Simple as that.

The various anti-evolutionist conferences I attended typically
showcased a bizarre mix of science and revival. The speaker might
make a scientific-sounding point about geology or whatever, and the
audience would reply with "Amen!" or "Praise God!" It was common
for speakers to open their presentations with prayers.

Anti-evolutionism is useful for whipping sympathetic crowds
into ecstasies of religious fervor, but it is utterly useless at getting
results in practical situations. Anti-evolutionists are much better at
propaganda than they are at solving problems, and that is the primary
reason scientists hold their ideas in such low regard.

2.7   NOTES AND FURTHER READING

The evidence for evolution has been laid out in many books and
    websites. Especially helpful in this regard are the website by Theobald
    (2012) and the books by Coyne (2009) and Dawkins (2009). Coyne and
    Dawkins wrote their books independently and published them at
    around the same time. It tells you something about the strength of the
    evidence that there is surprisingly little overlap between them. The
    short book by Rogers (2011) is also helpful. For a book-length
    treatment of the fossil evidence for evolution, see Prothero (2007).
    Prothero's more recent book (2020) is also a valuable reference in this
    regard. It presents some of the major lines of evidence for evolution by
    telling the stories of 25 major discoveries in the discipline's history.
    Isaak (2007) contains succinct refutations to hundreds of
    anti-evolutionist arguments.
The book by Mayfield (2013) provides a lengthy and eloquent discussion
    of the power of a variation/selection mechanism to produce complex
    results, not just in evolution, but in other fields of human endeavor as
    well. Appropriately, his book is titled *The Engine of Complexity*.

In my discussion of the evidence for evolution, I briefly mentioned the subject of vestigial structures. There is a common misconception that "vestigial" is a synonym for "nonfunctional." This is not correct, though it must be admitted that some biologists who have written on this topic have not always been as precise in their usage as they ought to have been. Anti-evolutionists seize on this misconception to argue that if a structure serves even the most minor purpose, then it is therefore not vestigial and not evidence for evolution. For example, they would object to my use of the human appendix as an example of a vestigial structure on the ground that recent research suggests it plays a minor role in strengthening the immune system. But this trivial function is entirely irrelevant to the judgment that the appendix is a vestige. Biologists typically define a vestigial structure to be one that exists in a reduced and rudimentary condition relative to the same complex structure in other organisms, and the human appendix fits that definition perfectly. Tasked with designing a structure that does for the immune system what the appendix is believed to do, no engineer would consider using a shriveled-up cow stomach.

The book by Shubin (2020) provides an accessible and up-to-date discussion of how paleontology and genetics, working together, are revealing the mechanics of some of the major transitions in evolution. Shubin's book will dispel any lingering sympathy you might have for the unbridgeable gaps argument. Shubin discusses many examples of evolutionary cooption. The article by McLennan (2008) also provides a readable discussion of this topic.

Schwab (2012) is a veritable encyclopedia of evidence for the evolution of eyes. The books by Land and Nilsson (2012) and Glaeser and Paulus (2015) are also valuable resources on the subject of eye evolution. The journal article by Gregory (2008) is a general reference on the evolution of complex organs that also includes a lengthy discussion of how eyes arose. These references are the justification for my claim that eye evolution is no longer considered to be especially mysterious. The older book by Dawkins (1996) is also helpful for thinking clearly about the evolution of complex structures.

There are many books discussing examples from anatomy and molecular biology that are incomprehensible as products of intelligent design, but which make perfect sense if you view them as the products of evolution. The books by Avise (2010), Hafer (2015), and Lents (2018) are good representatives of the genre.

The "senseless signs of history" argument is very powerful. For all the talk about how complex adaptations pose a challenge to current theory, when you look at how these systems are actually built, it is evolution that seems obvious and intelligent design that seems ridiculous. Anti-evolutionists have nothing cogent to say against this. Advocates for YEC at least confront the problem head on, gamely making the case that every instance of seemingly weird design is actually in some way the handiwork of a master engineer. Unfortunately, the weirdness is just too blatant and ubiquitous for this to be compelling. They also sometimes argue that some undesirable aspect of nature is a consequence of human sin, but discussing the merits of such explicitly religious theories is beyond the scope of this book.

For the ID perspective on the "senseless signs of history" argument, we can consider this quote, from biochemist and ID proponent Douglas Axe:

> Another way of downgrading life is to assume the role of a bio-critic – someone who looks for faults in the design of living things. As one example, the giant panda has a protruding bone in its wrist that serves a thumb-like role, enabling the bear to grasp bamboo. The fact that this bone (called a *radial sesamoid*) isn't a true jointed thumb like ours has led some people to view it as a makeshift adaptation that no good designer would employ. Not surprisingly, others argue that it *is* a good design. For my part, I find myself evaluating the people more than the panda. None of these people, however earnest they may be, have any deep grasp of the principles of design and development underlying sesamoid bones or thumbs, to say nothing of pandas. *(Axe 2016, 77)*

As we noted in Section 2.4, appealing to design constraints will not help the anti-evolutionists. The construction of the vas deferens or the recurrent laryngeal nerve is tantamount to placing a lamp right next to a wall outlet, and then attaching a hundred-foot extension cord before plugging it in. Do you really need a degree in engineering to understand why that is a bad idea? Complex adaptations are invariably jury-rigged and cobbled together from available parts, precisely as they would need to be for natural selection to be a viable hypothesis. Appealing vaguely to design constraints is not a serious response to this evidence.

Nor is it a serious response to protest that we cannot know the motives of the designer, who might have had inscrutable reasons of his own for designing things as he did. A design hypothesis can always be salvaged by this move, which is one reason biologists do not see design as a helpful concept in their work. The roads in Figure 2.1 might have been designed by a sadistic engineer who specifically wanted there to be crashes at that intersection, and the inefficient arrangement of pipes in my house might have been designed by an engineer who just wanted to use up some extra pipes he had lying around. These possibilities in no way mitigate the force of our argument. We are not claiming that bad design implies no design. Rather, we are making two separate claims. One is that weird kludges and inefficient design are the expected result of evolution by natural selection, which builds complexity by clumsily modifying existing structures. The other is that when human engineers design a system with a purpose in mind, they specifically try to avoid kludges and inefficiency. Therefore, when we observe that kludges and inefficiency are ubiquitous among biological adaptations, we should see that as strong evidence for evolution and against design.

The principles of "evolutionary game theory" have been laid out many times. The classic book by Maynard Smith (1982) remains relevant. The more recent book by Barash (2003) is a nontechnical overview of game theory, though only certain parts of the book are specifically about applications to biology. The anthology edited by Dugatkin and Reeve (1998) and the recent survey by McNamara and Leimer (2020) are representative of professional work in this area. As mentioned in this chapter, the tremendous success of game theory models in ethology is yet another line of evidence for evolution. The underlying mathematical models specifically assume the animal behaviors under investigation evolved by natural selection.

I mentioned applications of game theory in ethology, in the study of animal behavior. Game theory has also been applied successfully in the study of plant ecology. The paper by McNickle and Dybzinski (2013) is one of many references.

For discussions of the myriad ways evolutionary thinking routinely leads to progress in the daily work of scientists, I recommend the anthologies edited by Losos (2011) and Losos and Lenski (2016). For applications of evolution to medicine, have a look at Taylor (2015).

This is a book about evolution, which is separate from the origin of life. Evolution makes it possible to explain how a relatively simple sort of life billions of years ago can eventually transform into a very complex sort of life. This is no small accomplishment. However, evolution presupposes the existence of some sort of life, and therefore cannot be used to explain how life originated from nonlife. You will need a different theory to explain that, and that theory will have more to do with physics and chemistry than with biology. Anti-evolutionists often direct their fire at various theories for the origin of life, sometimes using essentially the same mathematical arguments they use against evolution. We will not discuss those arguments in this book, except to note that they are no more successful against the origin of life than they are against evolution. There is much that is unknown about the origin of life because it was a one-off event that happened billions of years ago under environmental conditions that were wildly different from anything we find today. But it is not completely mysterious either, with many partial results and numerous viable theories. For recent, accessible writing on this question, have a look at the books by Lane (2015), England (2020), and Prothero (2020, ch. 17).

3 The Parallel Tracks of Mathematical Reasoning

Having covered the basics of evolution, let us now consider the basics of mathematics.

When asked to describe what mathematics is, most people shudder and talk about mind-numbing arithmetical algorithms and tedious symbol manipulation. They have unpleasant flashbacks to high school algebra classes. Mathematicians find this frustrating since too many people profess their distaste for our subject without ever having experienced the real thing. Arithmetic and algebra are tools that we use in doing our job, but they are not what get us excited about our work.

As an analogy, woodworkers had better be proficient with tools like a band saw and a drill press. They need to be able to hammer a nail and drive a screw. But using saws and drills, hammers and screwdrivers, is not really the point of it all. For woodworkers, the pleasure comes from seeing a pile of wood, envisioning a finished project, and using their skills to bring it to fruition. Mathematicians experience something similar. For us, the pleasure comes from developing abstract models of reality and using them to learn something about how the world works.

My students often complain that mathematics is difficult because it is so abstract. I understand their frustration. Our inability to handle and manipulate mathematical objects can make it difficult to wrap our minds around them. However, there is an important sense in which my students have it backward. Mathematics is simple, it is reality that is complex.

Everyone who has ever looked at a road map understands this point. (Nowadays people typically rely on global positioning systems to get around, but I will assume everyone still understands this reference.) The reality is a dense network of roads, only a small portion of which is visible when you are lost among them. What is needed is a diagram to show you how all of the roads interrelate, and that is what the map provides. The map omits most of the reality. It does not depict the location of every squirrel and tree, and it does not show you every house or shop you will encounter. It is instead an abstract representation of those aspects of reality specifically relevant to getting you to your destination. The map is useful precisely because it is abstract.

You probably did not think you were doing mathematics by looking at a map, but in a very practical sense you were. Looking at an abstract representation of reality to plot your route is the same kind of activity as manipulating numbers to learn about physical objects. For example, if there are five kids over here and another three over there, then there are eight altogether. Something important has been accomplished upon realizing that this statement has nothing to do with kids, but is instead a statement about the abstract properties of collections of objects.

"Mathematical modeling" is the art of designing abstract representations of reality, with the intent of learning about reality by studying the abstractions. The basic idea is to discard most of reality so that the model is amenable to study, and then to hope that the remaining bits are the parts relevant to our problems. If our ultimate concern is with the real world, and if we see the model as primarily a tool for learning about the world, then we are doing science. If instead we study the abstract model for its own sake, then we are doing mathematics. This distinction is somewhat crude, and the lines between science and mathematics are often very fuzzy, but we have captured something important nonetheless.

Here is an example to show how the process plays out. In the eighteenth century, the city of Königsberg was located in Prussia. (Today it is known as Kaliningrad, and is located in Russia.) The city was divided into four land masses by the Pregel River, and these

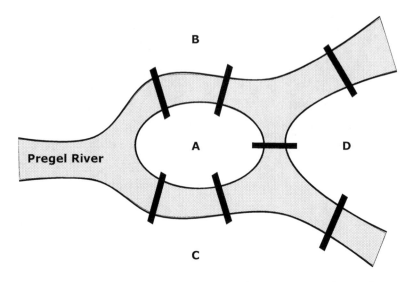

FIGURE 3.1 In the eighteenth century, the Prussian city of Königsberg was divided into four land masses, labeled *A–D*, by the Pregel River. The masses were connected by seven bridges, as shown.

land masses were connected by seven bridges, as shown schematically in Figure 3.1. The locals wondered whether it was possible to walk through the city in such a way that each bridge was crossed exactly once.

This problem came to the attention of Leonhard Euler (pronounced OY-ler), who was a mathematician of some prominence at that time. He devised an abstract model in which each land mass was represented by a dot, with the bridges represented by (possibly curved) lines connecting the dots. This is shown in Figure 3.2.

An arrangement of dots and lines of this sort is today referred to as a "graph." It is customary to refer to the dots as "vertices" and the lines as "edges." Since this particular graph allows multiple edges between the same pair of vertices, it is commonly referred to as a "multigraph." The number of edges coming out of a given vertex is commonly called the "degree" of that vertex.

Euler noticed that every vertex in his figure had an odd degree. Specifically, vertex A has degree 5, while vertices B, C, and D each

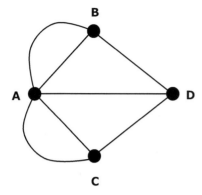

FIGURE 3.2 An abstract model
for the arrangement of land
masses and bridges in
Königsberg.

have degree 3. Imagine that you are walking along the graph. Each
time you first enter, and then leave, a vertex, you use up two available
edges. Now suppose there really was a walk that traversed each bridge
exactly once. Then, if we visit a vertex exactly k times during the
walk, then its degree must be $2k$, which is an even number. The only
exceptions to this are the starting and ending vertices, which might
have an odd degree. In other words, our walk will only be possible in a
graph in which no more than two vertices have odd degree. Since that
is not the case here, we see that the locals will search in vain for the
desired walk. A walk through a graph in which every edge is traversed
exactly once is today called an "Eulerian walk," in Euler's honor.

This is what modeling looks like when it is successful. We
devised an abstract model for a practical problem, and by studying the
model we found a solution to the problem. The model stripped away
all the distracting, irrelevant detail – such as the layout of roads on the
land masses or the differing demographics among them – and allowed
the focus instead to be on the really important detail – the evenness
or oddness of the number of bridges on each mass. (Mathematicians
would call this the "parity" of the number of bridges.)

The story does not end here. Once you have the idea of modeling something with a graph, you begin to see graphs everywhere.
Electrical engineers might see the vertices as representing circuit
connections and the edges as representing the wires that join them.
Physicists and chemists might see the vertices as representing atoms

in a molecule and the edges as representing bonds between them. In biology the vertices might represent proteins, with the edges representing interactions among them. Ecologists might see the vertices as habitats and the edges as animal migration routes.

If a particular abstract model can be used to represent many physical phenomena, a mathematician will argue that the abstraction is worth studying for its own sake. From this observation, the modern field of "graph theory" was born, and it remains an active area of research today. We might hope that anything we discover about graphs will simultaneously be useful to engineers studying circuits, to physicists and chemists studying molecules, to biologists studying protein interactions, and to ecologists studying animal migrations. Graph theory has indeed contributed to all of those fields.

Let us try a more complex example. A standard problem in physics is to predict the motion of a projectile. Perhaps we throw a baseball, for example, and we want to predict how high it will go and how long it will take to return to the ground. We know from experience that the ball will trace out a graceful arc, with the result looking something like Figure 3.3.

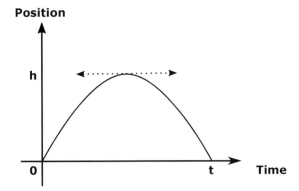

FIGURE 3.3 An abstract model for the trajectory of a thrown ball. We place time on the x-axis and position (or height) on the y-axis. The dotted line represents the maximum height attained by the ball; note that the tangent line is horizontal at that point.

You probably encountered diagrams of this sort in a high school math or science class. The diagram is called the "graph" of the function, and we should be clear that this is just an entirely different use of the word "graph" from our previous example. At any given time, the ball will be found at some height above the ground. It is convenient to place time on the x-axis and position (which is the same as height in this case) on the y-axis. We can then draw a curve that shows the ball's height above the ground at any point in time. Presumably we start our stopwatch at the moment the ball leaves our hand, and we can take the ball to be at ground level when we start the experiment. (We will not worry about details like the height of the person throwing the ball.)

A smooth curve of this sort is referred to as a "continuous function," and we can read off a lot of information about the path of the ball from this simple picture. For example, the maximum height occurs at the dotted line, and I have labeled that point h on the y-axis. The dotted line is called the "tangent line" to the curve at that point, and you will notice something interesting about the tangent line at the maximum height: It is perfectly flat. The tangent line at a point a little to the left of the maximum would be pointing up since the ball is still rising, and a little to the right it would be pointing down since the ball is now falling.

We could also note that the experiment ends at the moment when the ball returns to the ground, which corresponds to 0 on the y-axis. I have labeled that point with a t on the x-axis. Mathematicians refer to points like this, where the function crosses the x-axis, as the "roots" of the function.

The point is that if you want to understand the path of a projectile, you do not study projectiles. Instead you study continuous functions. Nor do you ask questions like, "How high will it go, and where will it land?" Instead you ask, "How do I find the roots of a continuous function?" or "How do I locate the points where the tangent line is perfectly flat?" The branch of mathematics that answers such questions is calculus.

Let us push this a little further, though I hasten to add that it will be unnecessary to understand every detail I present here and in the next few paragraphs.

After we throw the ball, why does its path change at all? Why does it not just leave your hand in a straight line and keep going until it flies off into space? The answer is that there is a gravitational force on the ball, pulling it down. The only other significant force is air resistance – the ball has to push the air molecules out of the way as it moves, and the molecules push back, so to speak. However, if the ball is reasonably dense, like a baseball or cannonball, then it does no harm to ignore air resistance.

In the language of physics, a force is something that imparts an acceleration, meaning that a force is something that changes your velocity. If gravity is the only force acting on the ball, then its velocity is changing in a constant manner. If we think in terms of graphs, then velocity is really the same thing as slope – both quantities can be seen as measuring the rate at which you are changing your position with time. We now ask, "What sort of function has a constant slope?" You might remember the answer to this question from an old algebra class: A straight (nonvertical) line. If you are walking up the line, the slope feels the same to you regardless of where on the line you find yourself. This is shown in Figure 3.4.

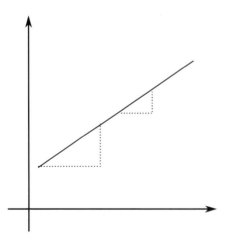

FIGURE 3.4 The only function with a constant slope is a straight line. No matter what two points we use, the slope always comes out the same, as illustrated by the dotted lines.

We now know that the velocity of the ball's motion is described by a straight line. We can relate this to the position of the ball by noting that, just as the acceleration describes the rate at which your velocity is changing, the velocity describes the rate at which your position is changing. In other words, velocity is the slope of position. We now ask, "What sort of function has a slope described by a straight line?" This one is harder than the earlier version of this question, so I will ask you to take my word for it that the answer is a parabola.

In this way, we can conclude that our ball will trace out a parabola after it leaves our hands, and with some additional hard work and algebra we can arrive at answers to our questions about the path of the projectile. Those details would be more tedious than enlightening, so let us move on.

3.2 YOU NEED BOTH RIGOR AND INTUITION

This section will be a bit technical, and I promise that you can just skim over the mathematical notation without losing the thread of the argument. However, since this book is largely about exposing bad mathematical arguments, I felt it important to provide a small taste of what real mathematics looks like.

If you pick up a high-level mathematics textbook and open to a random page, you are likely to find large amounts of incomprehensible symbolic notation. That portion of the page not given over to notation will be filled with equally incomprehensible jargon. If you are lucky, there might also be a diagram or two.

Attempting to read such a book feels much like reading something in a foreign language, and you can easily feel like you have no idea at all what is being asserted. You might even wonder why all of this technical detail is necessary. In Section 3.1, we managed to communicate some interesting mathematical ideas with a few pictures and normal language. Why must the textbooks be all but unreadable?

The reason is that mathematical objects are entirely abstract. They are not physical objects that can be handled and manipulated.

Consequently, they must be given precise definitions before we can presume to prove anything about them, and all of the jargon and notation allows a level of precision that is not possible using only everyday language. Let us see how this plays out with the objects discussed in Section 3.1.

Informally, a graph was a drawing composed of dots and lines, like the one shown in Figure 3.2. However, the drawing is really just a representation of a graph. It is a picture that helps us to visualize the relationships among whatever it is the vertices and edges represent, but it is not the graph itself. We need something more precise if we want to prove theorems about graphs.

I have in front of me a textbook on graph theory written by mathematician Robin Wilson. It offers the following definition of the term "graph":

> A **graph** G consists of a non-empty finite set $V(G)$ of elements
> called **vertices** and a finite family $E(G)$ of unordered pairs of (not
> necessarily distinct) elements of $V(G)$ called **edges**; the use of the
> word 'family' permits the existence of multiple edges. We call
> $V(G)$ the **vertex set** and $E(G)$ the **edge family** of G.
>
> *(Wilson 1996, 9)*

That sure seems a lot more complicated than what we discussed in the last section! Why did Wilson not just say, "A graph is a diagram in which you have a bunch of dots connected by a bunch of lines"? After all, that is pretty much how we defined things in Section 3.1.

The answer is that a definition solely in terms of dots and lines leaves too many unanswered questions. How many dots do you mean? Do we want finitely many dots, or are we allowing infinitely many? Are the lines allowed to start and end at the same dot? Do we want to think of the lines as arrows, so that there is a preferred direction in traveling from one dot to the next? Do we want to allow multiple lines to connect the same set of dots? In the problem of the Königsberg bridges, we did allow multiple lines because in some cases the same two land masses were connected by more than one bridge. On the other hand, if the dots represent cities and the lines represent the existence of a direct airplane flight between them, then maybe we only want to allow a single line.

Come to think of it, what do we even mean by dots and lines? In the brief list of applications we mentioned in Section 3.1, what we cared about was the relationship of what was paired with what. As we have discussed, the picture with the dots and the lines helped us to visualize and understand the graph. But just as the map is not the territory, the diagram is not the graph. If we erase the drawing from the page, the abstract relationships will still exist.

These are the reasons we needed a precise definition of the sort that Wilson provides. Notice that he anticipates and answers all of the questions I asked in the previous two paragraphs. He stipulates that we only want finitely many vertices. He emphasizes that the edges are unordered pairs, meaning that we are imagining lines and not arrows between the vertices. The phrase "not necessarily distinct" implies that it is fine for a line to start and end in the same place. (Such a line is called a "loop" in the graph.) And so on.

Our interest here is not really in the technical minutiae of how a graph is defined. We will not see graphs again after this chapter. However, I have belabored these points as an illustration of the parallel tracks of mathematical reasoning. Our examples have shown us two different ways of discussing mathematical concepts. On the one hand we have an intuitive understanding of our objects, of the sort presented in Section 3.1. I think of this as track one. On the other hand we have the rigorous, precise definitions with which mathematicians undertake their professional work. We shall call this track two.

As we proceed, we will often make distinctions between track one and track two mathematics. So let us be explicit about what we mean:

- TRACK ONE: Our intuitive understanding of what is really going on.
- TRACK TWO: The logically rigorous and precise technical definitions and theorems commonly found in high-level mathematics books.

Both tracks are critically important. If you try to get by solely with an intuitive, track one understanding, then it is too easy to make logical oversights when manipulating your abstract objects. Ultimately, everything you know about the objects is contained in their track two definitions, and that is why the utmost precision is necessary in formulating them. But the rigorous, track two definitions are sometimes so abstract that it is difficult to think clearly about the underlying objects. That is why you need track one as well.

For example, if you only think of a graph as an abstract set of vertices combined with an equally abstract set of edges, then you would probably never notice that problems about Eulerian walks are intimately related to the parity of the vertices. It is hard to notice that until you think in terms of dots and lines and literally imagine walking around the diagram using the edges as bridges between vertices. But if you only think in terms of dots and lines, then you will not be able to write down a convincing proof that Eulerian walks exist precisely when no more than two vertices have odd degree. That is why both tracks are necessary.

Let us push this just a little farther. We shall restrict our attention to so-called connected graphs, meaning that given any two vertices, it is always possible to walk from one to the other. We do not want a situation where one part of the graph is entirely cut off from some other part of the graph. It will also be convenient to consider Eulerian walks that start and end in the same place since that way we do not have to include the proviso that the first and last vertex are permitted to have an odd degree. To indicate that the starting and ending points are the same, we shall speak of an "Eulerian cycle" instead of an Eulerian walk.

Now, based on our experience with the Königsberg bridges, we have a good plausibility argument that Eulerian cycles through a graph exist precisely when all of the vertices have an even degree. How do we turn our plausibility argument into a proper proof?

First, we will need precise definitions of the terms involved. Continuing with Wilson's textbook, we shortly come to this:

Given a graph G, a **walk** in G is a finite sequence of edges of
the form $v_0 v_1, v_1 v_2, \ldots, v_{m-1} v_m$, also denoted by
$v_0 \to v_1 \to v_2 \to \cdots \to v_m$, in which any two consecutive edges
are adjacent or identical. ... A walk in which all the edges are
distinct is a **trail**. If, in addition, the vertices $v_0, v_1, \ldots, v_m$ are
distinct (except, possibly, $v_0 = v_m$), then the trail is a **path**. A path
or trail is **closed** if $v_0 = v_m$, and a closed path containing at least
one edge is a **cycle**.   (Wilson 1996, 26–27)

If you have no prior familiarity with the concept of a walk in a graph,
it would be hard to comprehend the concept based solely on this
definition. But if you already understand that you are just moving
from dot to dot along the lines of a diagram, it becomes much easier
to parse what this definition is saying.

After later stipulating that an Eulerian cycle is a cycle that contains every edge exactly once, and stipulating that a graph possessing
an Eulerian cycle is said simply to be Eulerian, we come to a properly
stated theorem:

Theorem (Euler 1736) *A connected graph G is Eulerian if and only if
the degree of each vertex of G is even.*

Since this is an especially famous theorem, credit is given to its
discoverer at the start of the theorem statement. The phrase "if and
only if" indicates that this is really two theorems in one: If a graph
is Eulerian then every vertex has even degree, and if every vertex has
even degree then the graph is Eulerian.

The first part seems simple enough, based on our earlier observation that a vertex that appears k times in an Eulerian cycle must
have degree $2k$. The other direction is surprisingly difficult, however,
and the next step would be to write down a logically cogent proof
showing it to be true. That level of detail is beyond anything we
need here.

We seem to have completely solved the Eulerian cycle problem:
A graph has an Eulerian cycle precisely when every vertex has even

degree. That is hardly the end of the story, however. A mathematician would now instinctively ask other questions. For example, what if we want to use every vertex instead of every edge? Can we walk around the graph using every vertex exactly once, starting and ending at the same vertex? This is called a Hamiltonian cycle, after the mathematician who first studied this question, and it turns out to be a much harder problem than the corresponding question for Eulerian cycles. There is nothing like our previous theorem in the case of Hamiltonian cycles. In other words, there is no easily observable property of a graph that tells you definitively whether or not a graph has a Hamiltonian cycle. Can we, then, prove theorems showing at least that certain specific kinds of graphs have Hamiltonian cycles? What if we ask the same questions for directed graphs, in which the edges have arrows on them indicating we can only traverse them in one direction? What if we have a weighted graph, meaning that there is a cost associated with crossing each edge? In such a situation, can we devise an algorithm that will allow us to walk around the graph at minimal cost?

Notice that we have left reality behind at this point. We are just asking questions about graphs and trying to answer them as best we can. This is what I mean when I refer to studying mathematical models for their own sake, separate from any real-world concerns. We take it for granted that graphs are worth studying since they model so many real-world phenomena, and therefore anything we learn about them will be at least potentially useful.

Let us go one more round to see how the two tracks play out with the notion of "continuous function." Again, let me emphasize that I am including the technical details simply to make a point. You can skim over them without losing the thread of our subsequent discussions.

Informally, the idea of "continuity" was that a projectile does not teleport from over here to over there. Instead it moves in a smooth, graceful arc. Sticking with track one, a continuous function is one whose graph can be drawn without lifting your pencil from the paper.

These are useful ways to picture what continuity means, but they again suffer from a lack of precision. If this is all we know about continuity, then what will we do when confronted with a function whose graph is too complex to draw? How can we come to understand continuous functions in general if we need a picture of each function individually just to know whether or not it is continuous?

We need a track two version of continuity, but in this case it is difficult to see how to devise a precise definition. We want to capture the idea that a continuous function has no breaks or jumps. If we imagine a small bug crawling along a curve, he should not encounter a point where he suddenly has to fall off a cliff, or jump a great height, to continue on his way. Roughly, we can say that a continuous function is one where the behavior of the function near a point matches the way the function behaves at the point. This is shown in Figure 3.5.

Formulating a proper definition of continuity therefore requires making a distinction between behavior at a point from behavior near a point. This distinction is captured by the notion of the "limit" of a function at a point. With a bit more work, we eventually arrive at something like this:

Definition 1 *We say that the limit of the function $f(x)$ at the point $x = c$ is L if for all $\epsilon > 0$, there exists a $\delta > 0$, such that if $|x - c| < \delta$, then $|f(x) - L| < \epsilon$. In this case we write*

$$\lim_{x \to c} f(x) = L.$$

The function $f(x)$ is continuous at $x = c$ if $\lim_{x \to c} f(x) = f(c)$.

It is the work of several class periods in a calculus course to convince students that these definitions really do capture our intuitive notion of continuity.

To parse the technical definition, keep in mind that the notation "$|x - c|$" can be understood to mean "the distance between x and c." The Greek letters δ (delta) and ϵ (epsilon) should be thought of as really small positive numbers. With that in mind, the definition of "limit" is really saying something like this: If x is very close to c, then $f(x)$

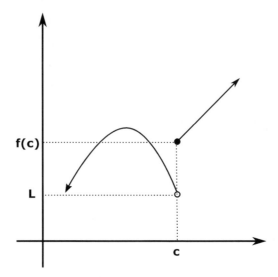

FIGURE 3.5 This function is discontinuous at the point c. A bug crawling over the top of the hill and heading toward c will expect to end up at the point L when he gets there, indicated by the open circle. But he is in for a surprise because the function suddenly jumps to a different value when we reach c on the x axis. In other words, the function's behavior near the point c does not match its behavior at the point c. To the left of c the function behaves one way, but then it does something completely different at and to the right of c. The arrows at the far left and far right of the diagram indicate that the curves continue infinitely in the indicated direction. They are separate from what is happening at the point c.

is very close to L. In terms of Figure 3.5, the bug crawling along the curve thinks he will end up at the open circle, and that represents the limit of the function at that point (at least as we approach from the left).

The definition of continuity then says that the value L that we are approaching should match the value of the function at the point. That is precisely what does *not* happen in Figure 3.5. The bug thinks he will end up at the open circle, but he is in for a surprise. The function value actually jumps at that point, and that is why the function is discontinuous there.

This sort of thing takes a lot of getting used to. You were probably much happier when a continuous function was just one that could be drawn without lifting your pencil from the paper. However, this is another nice illustration of the parallel tracks. If you have no inkling that a continuous function is something like a smooth, graceful arc, then it will be hard to understand what the technical definitions are getting at. But if you only think of continuous functions as smooth, graceful arcs, then you will be in real trouble when working with complicated functions whose graphs are not readily forthcoming. You need to work with both tracks simultaneously.

Though this is a book about mathematics, I have mostly tried to avoid using notation and jargon. Going forward, it will not be necessary to understand the symbols I have used in this section. However, I did have a reason for showing them to you.

When you are trying to communicate the main ideas of a branch of mathematics to a lay audience, it is perfectly fine to employ a casual, informal, track one understanding of the concepts. But if you claim that you have mathematical proof that a major branch of science is thoroughly rotten to its core, then you had better be ready to work at a track two level. Do you *really* have a strong mathematical argument, or are you just aping mathematical terminology to create a phony air of precision? Formulating a good mathematical argument requires meticulous attention to detail and the utmost clarity in defining your model.

As we proceed, we shall see that anti-evolutionist mathematics is plagued by an inability to get both tracks right. Inevitably, one or the other of the tracks is missing.

Sometimes they make audacious claims to have found a mathematical disproof of evolutionary theory, but then present only a muddled, track one argument. When you ask for the track two details that would make the argument persuasive, you find that nothing is forthcoming. Other times they present copious track two minutiae, but when you try to work out the track one understanding of what is really going on, you find that the jargon and notation are nothing but gobbledygook, and that nothing substantive is really being said at all.

We will see examples of both fallacies in the chapters to come.

3.3   BAD MATHEMATICAL MODELING

There is an old joke about a group of dairy farmers who want to increase milk production from their cows. They bring the problem to some local physicists, who spend the next week working on it. The lead physicist then reports their results to the farmers. He says, "We have a solution to your problem, but it only works if you assume spherical cows in a vacuum."

There are several aphorisms in science that make the same point. It is sometimes said that when presuming to devise a scientific model, you should make everything as simple as possible, but no simpler. Another holds that all models are wrong, but some are useful. The point is that all models are based on unrealistic simplifying assumptions, but sometimes those assumptions are near enough to being true that we never notice the difference.

For example, consider our model for the path of a ball from Section 3.1. Our model was wrong in the sense that we ignored air resistance. Even for dense objects like cannonballs or baseballs, air resistance has an effect and therefore our model inevitably gives the wrong answer. However, the model was useful because the difference between what the model predicts and what actually happens is so small that we would need very sensitive equipment to measure the difference. Had we used something less dense, such as a table tennis ball, it would be a more serious mistake to ignore air resistance.

The history of science records many instances of clever modeling being undone by invalid assumptions. A famous example of particular relevance to us involves a nineteenth-century attempt to estimate the age of the earth. Physicist William Thomson, who would later be known as Lord Kelvin, published an estimate for the age of the Earth. To carry out his calculation, Kelvin imagined that the Earth began as a molten ball comparable to the Sun. He then estimated how much time would be needed for the Earth, having started in such a state, to cool to its then-current surface temperature. On this basis, he concluded that the Earth was somewhere between 20 million and 400 million years old.

Darwin saw these estimates as a stumbling block for his theory, since the evolutionary process needs a very long time to unfold. However, subsequent discoveries showed that Kelvin's estimates were substantially off because of faulty assumptions. Earth's surface is warmed by two sources of energy that played no role in Kelvin's calculation. One source is radioactivity since radioactive elements in the Earth's crust emit heat as they decay. The other is convection currents within the Earth's mantle that cause cooler material near the surface to fall toward the center of the Earth, while warmer material from closer to the Earth's core rises to the surface. Kelvin's estimate for the rate of cooling at the Earth's surface was a bad overestimate, and therefore his estimate for the Earth's age was a comparably bad underestimate. Kelvin considered a simplified model, but it turned out to be too simple.

A more dramatic example comes from mathematical estimates in the 1920s purporting to show that it was impossible for a person to run a four-minute mile. These estimates were based on solid data about the rate at which a person could take in oxygen and the rate at which energy could be released. However, that *something* was wrong was made clear in 1954, when British athlete Roger Bannister became the first person to run a mile in less than four minutes.

This story is reminiscent of the urban legend that scientists at one time claimed to have proved that a bumblebee could not fly. The small nugget of truth underlying this story is that certain simplistic models of bumblebee flight, specifically ones that assume bees possess a fixed wing like an airplane as opposed to one that flaps like a bird, show that a bumblebee is too heavy and nonaerodynamic to fly. But no one ever used that as evidence that bumblebees could not fly, in defiance of the manifest fact that bumblebees are routinely seen to fly. Instead it was taken as evidence that the models were too simplistic.

These last two examples illustrate a point to which we will return frequently in the pages ahead. The point is this: If copious physical evidence suggests that something has occurred, but someone's mathematical model says the thing is impossible, then it is probably the model that is wrong.

This is especially relevant in discussions of evolution and creationism. As described in Section 2.2, we have copious physical evidence in support both of common descent and of the ability of natural selection to explain the appearance of complex adaptations. Furthermore, each of the individual steps of the evolutionary process is known to occur: Genes really do mutate, sometimes leading to new functionalities, and natural selection really can string together several mutations into adaptive change. On a small scale this has all been observed. If small amounts of evolutionary change are seen to occur over short time scales, then it is hard to see how some abstract principle of mathematics will rule out large amounts of change over longer time scales. After all, if we have a mechanism for small-scale change, then we should be able to apply that mechanism over and over again to produce large-scale change.

In other words, copious physical evidence strongly suggests that evolution has occurred, but anti-evolutionists claim to have mathematical models showing this to be impossible. Given this evidence, and given the track record of evolution in producing results when applied to practical problems, our instinctive reaction should be to suspect the models are too simplistic and to go looking for the dubious assumptions underlying them.

As we shall see, we will never have to look too hard.

3.4   ANTI-EVOLUTIONISM'S FALLACIOUS MODEL

The anti-evolutionists have a model of their own, one that underlies most of their argumentation. In true evolutionary style, we shall build up to their model gradually.

You are probably familiar with "word search" puzzles, in which meaningful words are buried in an array of meaningless letters. When

I was growing up, my father used to make puzzles like this for me to solve. He would lay them out on a standard sheet of graph paper, placing one letter in each cell. Sometimes these puzzles were as big as 30×30 cells. With 900 cells to search through, it could be surprisingly difficult to find the small islands of meaning within the larger ocean of meaninglessness. Especially at age six.

My father was skillful in constructing these puzzles. When he filled in the camouflage letters, he did it so that the distribution of letters in the array was comparable to the distribution of letters in English words generally. It would not do to have have the background consist entirely of Xs and Zs since meaningful words would then stand out clearly. He did this by imagining coherent, but nonsensical, sentences in his head and filling in the boxes accordingly, spelling out the sentence as a sequence of letters and placing those letters in randomly chosen empty cells. And though he always provided a word list beneath the array, organized around a theme like animals or US states, I knew there would always be a few bonus words as well. My name was always in the array, as was my older brother's name, Neil. "Mommy" and "Daddy" were also always in there somewhere.

Even as a child, I quickly realized that picking letters at random and then scanning in all eight directions was terribly inefficient. Sometimes I would get lucky with that approach and just happen to land on the first letter of a word, but more often I was wasting my time. A far better approach, I discovered, was to look for pairs of letters. For example, if I was looking for "Rhode Island," I would scan the array looking for adjacent Rs and Hs. If the word had a Q, then I knew I only had to look in directions where the next letter was a U. Unusual letters like X and Z were big giveaways, since typically there were only a few of them in the array.

I have belabored these reminiscences because the puzzles themselves, and my approach to solving them, are strongly analogous to the way anti-evolutionists see evolution.

As they see it, evolution also involves searching for small targets in large spaces. In the puzzles, the solver looks for a small number of meaningful words against a large backdrop of meaningless letters. In evolution, nature searches for functional genes against a backdrop of theoretically possible, but ultimately nonfunctional, genetic sequences. The person solving the puzzle is unlikely to be successful by searching randomly. Likewise, the evolutionary process has little hope of finding functional genes if it searches randomly through the space of all possible genes. And just as the puzzle-solver must employ intelligence to find efficient algorithms for searching the space effectively, so too does the evolutionary process need intelligent guidance if it is ever to have any success.

Let us make this more precise. Biologists use the term "genotype" to refer to an organism's complete set of heritable genes. This term should be contrasted with "phenotype," which refers to all of the observable characteristics of the actual organism. We can say, for example, that the phenotype is the end result of an interaction between the genotype and its environment. Though it is somewhat simplistic, genes can be usefully thought of as sequences of letters drawn from an alphabet of four possibilities. Looking at things in this way, we can imagine a vast "genotype space" consisting of all possible sequences of the four letters up to a given length. A few of those sequences will represent genes which, were they to be found in an organism's body, would perform some useful function. However, most of those sequences will not represent functional genes. Since genes code for proteins, we could as easily think of "protein space" instead of "genotype space."

As an analogy, we can imagine a large space consisting of all possible sequences of no more than 10 letters drawn from the standard set of 26. It turns out that there are more than 141 trillion such sequences. Among those sequences will be all English words of no more than ten letters, but most letter sequences will just be gibberish.

Returning to the metaphor of "genotype space," a given animal's genotype can then be viewed as one point in the vast space of all possible genotypes. Each of that animal's offspring represent other points in genotype space that are near the parent's points. They represent different points, because offspring are never genetically identical to the parents, but they are nearby, because they were obtained by small mutational changes to the parents' genome. As we go down through the generations, we can imagine tracing out paths in genotype space.

We can take humans to represent one point in genotype space and lobsters to represent another. Presumably, they represent points that are very far apart. But evolution implies that if we go back through the generations leading to humans, and back through the generations leading to lobsters, then these paths will eventually converge. In fact, according to evolutionary theory, we can make the same claim for any two modern species, no matter how different they appear. If evolutionary theory is correct, then genotype space must be structured in such a way that it is possible to traverse vast distances by small mutational steps, aided by natural selection.

Anti-evolutionists claim that the space is not structured like this and that arguing otherwise is tantamount to believing that a vast word search can be solved by random trial and error. They assert that genotype space is so vast, and that functional sequences are so rare within it, that it is not feasible for natural causes to trace out paths connecting vastly different life forms. In particular, they claim that mutation and natural selection cannot trace out paths leading to modern, complex adaptations.

Biologists do not agree with this conclusion. They point to the copious empirical evidence and numerous practical successes of evolutionary theory, and then argue, in effect, "It did happen, therefore it can happen." If you try to show them otherwise based on an abstract model, they will respond, appropriately, in the manner of the previous section. They will say, "I'm sure it's a lovely model. But if physical evidence says one thing and an abstract model says something different, then we should ignore the model and not the evidence." Then they will get back to work as though nothing had happened.

This means the anti-evolutionists will need to support their claims with a strong argument, and this is where the mathematics comes in. We can summarize their basic strategy like this:

(1) Model evolution as a search.
(2) Represent complex biological structures as targets of the search.
(3) Invoke some piece of mathematics meant to establish the extreme implausibility of evolution finding the target.

The exact space to be searched depends on the author. Sometimes they refer to the space of all possible proteins, instead of to the space of all possible genotypes. In other cases they might have in mind phenotypic structures like eyes or wings. However, the precise space is irrelevant to the logic of the argument.

Several branches of mathematics have been deployed in the service of item three. However, in modern anti-evolutionary discourse the argument always comes down to one of the following two methods:

(3a) Carry out a calculation to establish that the probability of the target is too small.
(3b) Produce a general principle or theorem to conclude that evolution could not find the target.

The remainder of this book will examine the specific manifestations of methods (3a) and (3b) in the anti-evolutionist literature. We will find all of them to be seriously wanting. However, it will be useful to enter into this discussion with a general understanding of the kinds of things that go wrong with arguments of this sort.

Mathematical anti-evolutionism inevitably fails for at least one of the following two general reasons:

(3a') The calculation described in (3a) is based on an unreasonable reduction of probability to combinatorics, thereby rendering it meaningless.
(3b') The general principle or theorem mentioned in (3b) is irrelevant to assessing evolution's fundamental soundness.

Note that "combinatorics" is the branch of mathematics devoted to counting arrangements of things, but we will defer further discussion until Chapter 5.

I say "inevitably" because as a practical matter it is impossible to develop the search model with sufficient detail to draw persuasive mathematical conclusions from it, at least not of the sort the anti-evolutionists desire. Let me explain why I say that.

When mathematicians refer to a "space," they usually mean a collection of objects that have certain relationships to one another. There are two sorts of relationships that are especially relevant when we think about protein space. (The same considerations apply to genotype space, but it will be convenient to restrict our attention to just one space.)

The first important relationship is that some proteins are close to each other in the space while others are far apart. In other words, we have a way of measuring the distance between any two points in the space. Informally, if two long proteins differ in only one amino acid, then I can say they are close together in the space, and if they differ in many of their amino acids, then they are far apart. Since geometry is the branch of mathematics that studies the arrangement of points in a space with respect to one another, I will refer to these distance relations among proteins as the "geometric structure" of the space.

The second important relationship is that starting from a specific protein, genetic mutations are more likely to move us to nearby proteins than they are to faraway proteins. Moreover, some proteins are highly unlikely ever to be found in an organism that survives to reproductive age because they are physically harmful to those that possess them. Since probability is the branch of mathematics that attempts to quantify how likely or unlikely we believe something to be, I will refer to these likelihood relations as the "probabilistic structure" of the space.

Now, if the search model is to be developed to the point where grand mathematical conclusions can be drawn, it is clear that we need a detailed understanding of both the probabilistic and geometrical structures of the space. In other words, we need to be able to answer the following kinds of questions:

(4) If an organism is at a certain point in the space, how likely is it that its descendants will be able to reach other, distant points? Specifically, how

likely is it that a primitive sort of life lacking eyes can find a path through the space leading to organisms with eyes?

(5) How are functional proteins, viewed as points in the space, situated with respect to one another? Specifically, are functional proteins just tiny islands of functionality in an ocean of useless junk, or are they situated in ways that can serve as stepping-stones to ever more complex adaptations?

The anti-evolutionists claim to be able to answer these questions with mathematical precision and, furthermore, claim that the answers show that evolution is fundamentally unsound as a scientific theory.

In reality, however, we have nothing like the information about this space that we would need to arrive at such conclusions. That is why mathematical anti-evolutionism inevitably fails. The remainder of this book will show why this is so.

3.5   NOTES AND FURTHER READING

The short book by Gowers (2002) is an excellent and highly readable account of mathematical thought generally. It is a useful remedy for anyone who thinks mathematics is just about arithmetic and symbol manipulation.

My description of Euler's work on the Bridges of Königsberg was somewhat oversimplified. Euler did not actually define a graph in the way we use that term today, but his work leads so naturally to that approach that I did not feel it was important to belabor the difference. For a thorough account of what Euler actually did, I recommend the article by Hopkins and Wilson (2004).

There are many legitimate uses of mathematics in evolutionary biology. Biologist Sergey Gavrilets writes:

> Since the time of the Modern Synthesis, evolutionary biology has arguably remained one of the most mathematized branches of the life sciences, in which mathematical models and methods continuously guide empirical research, provide tools for testing hypotheses, explain complex interactions between multiple evolutionary factors, train biological intuition, identify crucial parameters and factors, evaluate relevant temporal and spatial scales, and point to the gaps in biological knowledge, as well as

> provide simple and intuitive tools and metaphors for thinking
> about complex phenomena.   *(Gavrilets 2010, 46)*

For a survey of mathematical techniques used in evolutionary
analysis, try the book by Nowak (2006). Interestingly, Nowak includes
a chapter discussing applications of graph theory in evolutionary
biology.

Mathematicians typically use the term "metric space" to describe an
abstract set of points on which some notion of "distance" has been
defined. In this sense, it might have been more mathematically
correct to say that protein space has a metric structure, rather than a
geometric structure. However, I chose the latter term since "metric
space" is an expression that is mostly unknown outside of
mathematics, while most readers will recall that geometry studies
arrangements of points with respect to one another (with circles,
triangles, and lines, for example, being especially interesting
arrangements).

I suspect many biologists will object to the whole idea of modeling
evolution as a search. In everyday usage, carrying out a search
typically implies planning, foresight, and a pre-set target. For example,
a firm might search for a candidate to fill a position. The firm knows
the kind of person they are looking for, and they will organize the
search so as to maximize their chances of finding someone appropriate
to the job. None of this is analogous to evolution, which does not so
much search protein or genotype space as it does meander aimlessly
around it with no target in mind. Moreover, in normal usage we
assume that the space to be searched, and our standards for assessing
the desirability of the points within the space, typically referred to as
the "fitness landscape," remains static. Evolution is not like that. In
evolution we have numerous populations meandering around
different parts of the space, and their movements constantly change
the fitness landscape. These are salient points, and they certainly
show that modeling evolution as a search is deeply problematic. That
said, in discussing mathematical anti-evolutionism I am willing to
accept the search metaphor just for the sake of argument. The major
errors in their discourse have little to do with whether or not we
think this is a helpful metaphor, and I have no wish to get bogged
down in semantic disputes. We will revisit this point in Section 6.8.

4 The Legacy of the Wistar Conference

4.1 AGAIN, DOES EVOLUTION HAVE A MATH PROBLEM?

Biology is less mathematical than other branches of the physical sciences, and this has sometimes been used as a cudgel against it. The argument goes like this: While we can all agree that dissecting animals and classifying them into groups is important work, it is hardly the same as predicting the mass of an electron to fourteen decimal places or to working out the motions of the planets centuries in advance. Evolution is biology's premier theory, but it is not defended in the usual manner, is it? *Real* science uses equations to make precise predictions that are then verified by experiment. Evolution just points to a few fossils or anatomical comparisons and calls it a day.

So goes the argument, at any rate.

This attitude is far less prevalent today than it used to be. Physics is heavily mathematical because it studies simple, predictable objects like atoms and billiard balls. This makes it easy to capture their behavior in a few equations. Questions about complex, unpredictable, living things are less amenable to mathematical analysis. Mathematics is a useful tool for science, but it is silly to make mathematical precision the sole standard by which we evaluate a discipline.

Still, it is an occupational hazard for biologists to be condescended to by more mathematically-inclined scientists. It happens periodically that some nonbiologist turns his attention to the mysteries of life and, having thought about the problem for a few minutes, presumes to lecture biologists on how to do their jobs. Such people have seldom studied an actual animal, but they typically have some facility with symbol manipulation, and that is deemed sufficient to hold forth. The biologists, for their part, are expected to be grateful for this attention.

This attitude was on full display at a 1966 symposium held at the Wistar Institute in Philadelphia. The proceedings were published the following year under the title *Mathematical Challenges to the Neo-Darwinian Interpretation of Evolution*. The proceedings make for interesting reading since they not only contain transcripts of the various presentations, but also transcripts of the ensuing discussions.

The title is misleading because, of the seven main presentations, only two posed challenges to evolution's fundamental soundness. The first challenge came from Murray Eden, then an engineer at the Massachusetts Institute of Technology in the United States. The other was from Marcel-Paul Schützenberger, a mathematician then at the University of Paris in France. The other presentations had more to do with trying to formulate evolutionary questions in mathematical terms than they did with trying to refute the theory altogether.

Eden and Schützenberger were both eminent in their own fields, which lent an air of gravitas to their arguments. When the local preacher claims to have a decisive argument against evolutionary theory, the challenge is easily ignored. But when the likes of Eden and Schützenberger have something to say, they get a full hearing.

The symposium was chaired by Nobel laureate Peter Medawar. Among the presenters were Ernst Mayr and Richard Lewontin, both among the first tier of biologists of their time, and mathematician Stanislaw Ulam, who had been part of the famed Manhattan Project. The other attendees were likewise distinguished in their areas of expertise. In other words, there was some serious talent in the symposium's audience, and all involved were taking the issues *very* seriously.

Which made the actual presentations rather anti-climactic. For all of their eminence in their own fields, the anti-evolution arguments presented by Eden and Schützenberger were just bad, even given what was known back in 1966. That notwithstanding, they established the template followed by modern purveyors of mathematical anti-evolutionism. Digging into the minutiae of their claims will help us to understand the errors of their modern-day followers.

4.2   NATURAL SELECTION IS LIKE A TRUFFLE HOG

We begin with Eden's presentation. He summarized his argument like this:

> Any principal criticism of current thoughts on evolutionary theory is directed to the strong use of the notion of "randomness" in selection. ... The issue of plausibility is central to my argument; namely that when reasonable assumptions are made concerning certain natural processes, together with the assumption of certain specific kinds of randomness in the variation of heritable properties, then other phenomena which are empirically observable appear to be highly unlikely events.
>
> *(Moorhead and Kaplan 1967, 5)*

His main argument in defense of this claim involves the problem of finding functional proteins within the vast space of theoretical possibilities. A protein can be viewed as a long chain, each of whose links is one of twenty amino acids. If we imagine building a protein by randomly selecting two of these amino acids, then there are $20 \times 20 = 400$ possibilities. If instead we randomly selected three, there are $20 \times 20 \times 20 = 8,000$ possibilities. Most proteins are much longer than this, so Eden asked everyone to imagine a chain with 250 links. We find that there are 20^{250} possibilities, which is roughly 10^{325}. That is a one followed by 325 zeroes. For a comparison, most people would consider a billion to be a big number, but that only has nine zeroes. We can say, therefore, that the space of all possible proteins is monstrously large.

In contrast to this very large number, Eden put forth 10^{52} as the number of proteins that could ever have existed in any organism in all of natural history. This number was based on a crude calculation, the details of which are not important. By itself, 10^{52} is a large number, but it is very small compared to 10^{325}. Somehow, the evolutionary process has found a tiny set of functional proteins within a vast space of possibilities. This needs to be explained, and Eden suggested two possible solutions:

> Either functionally useful proteins are very common in this space so that almost any polypeptide one is likely to find has a useful function to perform or else the topology appropriate to this protein space is an important feature of the exploration; that is, there exist certain strong regularities for finding useful paths through this space. *(Moorhead and Kaplan 1967, 7)*

In the discussion after his presentation, Eden elaborated on this point:

> There is some path by which we have arrived at this relatively small corner in this large space, on the basis of a relatively small number of generations. What I am claiming is simply that without some constraint on the notion of random variation, in either the properties of the organism or the sequence of DNA, there is no particular reason to expect that we could have gotten any kind of visible form other than nonsense. *(Moorhead and Kaplan 1967, 14)*

An analogy will make his reasoning easier to follow.

There is a small fungus known as a truffle that many gourmets regard as a delicacy. Truffles are very expensive precisely because they are so rare and coveted. The problem is that they grow in forests underground, where they are hard to find. If you enter a forest at random and start digging, you are unlikely to find one. For this reason, truffle harvesters typically use trained pigs, called truffle hogs, to sniff them out. In so doing, they substantially increase their chances of success.

The point is that when you try to find small targets in large spaces, you need some kind of edge. In Eden's argument, proteins are playing the role of the truffles, and the large space of theoretical possibilities is playing the role of the forest. Eden is asking what plays the role of the pig.

We can respond to this request with another analogy. Imagine we are walking through a dense forest with a single, narrow walking trail. Suddenly we emerge in a clearing to find ourselves standing at the bank of a river that we need to cross. Since we lack a canoe or any similar conveyance, we go looking for a different solution. We notice a large rock we can use as a stepping-stone, so we jump to it. From there we notice another stepping-stone, and then a third. In this manner we are able to cross to the other side.

Upon reaching the other side, a local approaches us. He says, "That's amazing! I've been up and down this river many times and I happen to know that these are the only stepping-stones to be found anywhere along its entire length. You found the one, lonely, sequence of stepping-stones in a vast river. How on Earth did you do it?"

Continuing with the analogy, it is not hard to answer the local's question. We did not randomly search the entire river looking for stones. Instead we carried out a series of local searches in the neighborhood of wherever we happened to be. We were successful because, while stepping-stones are rare in the river as a whole, they were common in the small part of the river we needed to search.

The local might not be impressed by that response. He might say, "That is all well and good, but isn't it remarkable that you ended up right in a clearing from which you could reach the first stone. Had you started from anywhere else along the river, you would have been well and truly stuck. But you found the only possible starting place. Amazing!"

To which we would reply, "Still, there is no mystery. There was only one walking path through the forest, and it channeled us right to that starting point. We did not choose our starting point at random, but rather were forced there by the territory surrounding the river." And if the local now insists that we explain why the territory is so ordered as to lead us to that starting point, we would probably reply, "That's an interesting question, but it's entirely separate from the one we started with. There's no mystery about how I found the stepping-stones across the river, because the forest forced me to a point from

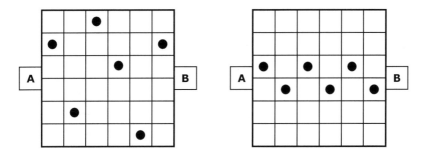

FIGURE 4.1 The grid represents a large space of possibilities, while the black dots represent useful stepping-stones connecting A to B. The stones are equally rare in both cases. On the left, the stones are too scattered to be useful, but on the right we can cross from A to B.

which I could find the first stone, and from there I only had to search small portions of the river at each step. Answering broader questions about why things are laid out like this requires experts in the history of this region, and I will leave it to them to answer your question."

The biological analogs are clear. If we want to understand how it was possible for protein A to later evolve into protein B, it is irrelevant that they live in a vast space of theoretically possible proteins, and it is irrelevant that functional proteins are generally rare in the space. What matters is whether functional proteins are sufficiently common in the vicinity of A and B to be used as stepping-stones. This is illustrated in Figure 4.1.

In fact, the general rarity of functional proteins in the space could even make it easier to find the path from A to B. To see why, let us return to my analogy. As I presented things, there was only one path of stones across the river. Consequently, no matter where we were, it was easy to find the uniquely correct next step since anything else would have landed us in the water. But suppose that instead of just one path there were actually many paths, only one of which takes us to the other side. The others just peter out into dead ends in the middle of the river. Also suppose the river to be so wide that we cannot see the other side until we have already traversed a large number of stones. In this scenario, we would likely explore a large number of dead ends before stumbling on to the correct path. We might even fall into the water just out of sheer exhaustion. It is not always advantageous to be confronted with many choices.

In biology, it is natural selection that ensures we remain on the correct path, so to speak. Each of an animal's offspring represents an experiment with a point in the space nearby to the point represented by the parent. If any of those offspring stray into the realm of nonfunctional proteins, then natural selection will step in to ensure their genes do not spread. The sheer rarity of functional proteins will ensure that the selection pressure to stay on the correct path is very powerful indeed.

Of course, a skeptic might reply to this in precisely the manner of the local in my story. He might wonder where protein A came from in the first place. The ensuing discussion would quickly trace the chain of causation back to the origin of life. At that point we would reply precisely as in the analogy, saying, "The origin of life is an interesting and important question, but it's just completely separate from the problem we started with. There is no mystery to how evolution finds small sets of functional proteins in a vast sea of gibberish because it was forced to a certain starting point by the origin of life, and from there it only undertook a series of local searches, ignoring the rest of the space. The origin of life has far more to do with physics and chemistry than it does with evolution, and we will defer to the relevant experts to answer that question."

In Section 3.4, I mentioned the need to understand both the geometric and probabilistic structures of protein space. Eden's argument provides our first opportunity to flesh out what that means. He based his argument largely on the vastness of protein space as a whole compared with the relative smallness of the set of proteins that are actually used. He then suggested that some constraints were necessary for evolution to be able to search effectively.

It is precisely the geometric and probabilistic structures that provide those constraints. The geometric structure relates to how functional proteins are situated with respect to one another: if they are arranged like stepping-stones then it does not matter that they are rare in the space as a whole. The probabilistic structure relates to the role of natural selection in the process: selection dramatically raises the probability of finding functional structures as compared to searching blindly.

At several points in his presentation, Eden briefly mentions that selection is part of the process, but he never seems to realize the importance of this fact. This failure was noted by other conference participants. Most notably, the aforementioned Stanislaw Ulam made the following comment at the start of his own presentation:

> [I] believe that the comments of Professor Eden, in the first five minutes of his talk at least, refer to a random construction of such molecules and even those of us who are in the majority here, the non-mathematicians, realize that this is not the problem at all.
>
> A mathematical treatment of evolution, if it is to be formulated at all, no matter how crudely, must include the mechanism of the advantages that single mutations bring about and the process of how these advantages, no matter how slight, serve to sieve out parts of the population, which then get additional advantages. It is this process of selection which might produce the more complicated organisms that exist today.
>
> *(Moorhead and Kaplan 1967, 21)*

This is well said, and it largely vitiates Eden's argument.

4.3   GENETICS IS DIFFERENT FROM COMPUTER SCIENCE

While Eden was relatively understated in making his case, Schützen-berger was far more assertive. He opened by saying, "Our thesis is that neo-Darwinism cannot explain the main phenomena of evolution on the basis of standard physico-chemistry," and he closed by saying, "[W]e believe that there is a considerable gap in the neo-Darwinian theory of evolution, and we believe this gap to be of such a nature

that it cannot be bridged within the current conception of biology."
(Moorhead and Kaplan 1967, 73)

If you are going to talk like *that*, you had better have a mighty good argument to back it up!

Schützenberger based his conclusions on two observations. The first was that evolutionary theory posited a connection between what he called "typographic" changes in genotypes and observable features of phenotypes. The term "typographic" is metaphorical in this context. We certainly are not thinking about a literal typewriter. Rather, Schützenberger's intent was that we are thinking about a genotype as a sequence of letters, and that mutations can be thought of as changes to those letters. The second was that when random typographic changes were made in computer programs, the result was usually an entirely nonfunctional program. He presented his challenge as follows:

> According to molecular biology, we have a space of objects (genotypes) endowed with nothing more than typographic topology. These objects correspond (by individual development) with the members of a second space having another topology (that of concrete physico-chemical systems in the real world). Neo-Darwinism asserts that it is conceivable that without anything further, selection based upon the structure of the second space brings a statistically adapted drift when random changes are performed in the first space in accordance with its own structure.
>
> We believe that it is not conceivable. In fact, if we try to simulate such a situation by making changes randomly at the typographic level (by letters or by blocks, the size of the unit does not really matter), on computer programs we find that we have no chance (i.e. less than $1/10^{1000}$) even to see what the modified program would compute: it just jams.
>
> *(Moorhead and Kaplan 1967, 74–75)*

To see what Schützenberger had in mind, imagine a recipe for chocolate chip cookies. Over here we have printed instructions telling us what to do. By carrying out the steps of the recipe, we end up with the actual cookies to be eaten. If we make random changes to the recipe's instructions, we are likely to produce no edible cookie at all, or at best a vastly inferior cookie.

Likewise for the blueprints of a building. The blueprints can be seen as instructions for assembling the building. As with the cookie, we have assembly instructions on the one hand, and a finished object on the other. Random changes to the instructions are likely to have a deleterious effect on the finished building, to put it mildly.

And likewise for Schützenberger's computer programs. You have coded instructions on the one hand, and whatever the program does on the other. Random changes to the code nearly always produce something worthless. Not only do you not get a functional program, you do not even get anything meaningful.

But according to evolutionary theory, the argument continues, random changes to genetic instructions somehow lead to meaningful change at the level of actual organisms. This is what is said to be inconceivable.

In the discussion following his talk, Schützenberger said this:

[I]n order to mediate between the space of chains of amino acids and the real world of organisms, some new construct has to be introduced, and principles have to be stated explicitly explaining how this mediation is conceivable.

At the level of molecular biology, we are told that we have a reasonably complete description of the mechanisms. Also, physiology is providing us with an understanding of organs. However, everybody seems to take for granted that there is no gap in between. I am not discussing the adequacy of each of the two extremes. I just point out that nobody seems to be able to give reasons why they have anything to do with each other. If there were explicit general principles, then we should be able to simulate something analogous, and we would have a lot of fun studying mathematical models showing the passage from disorder to order. *(Moorhead and Kaplan 1967, 76)*

Schützenberger's presentation is muddled, and it is not always clear precisely what he is asserting. In some places he seems to be making the modest claim that the mechanisms of development, through which the genotype is transformed into the phenotype, were very poorly understood at the time of the conference. Elsewhere, however, he seems to be making the much stronger claim that it was not even possible in principle for the genotype to determine the phenotype. For example, here is an exchange between Schützenberger and Ulam in the ensuing discussion:

> ULAM: My impression is that what you have said so far is that one does not understand now how the blueprint determines the existing physical objects. That, of course, the Darwinians or neo-Darwinians would freely admit. Now, the assertion that such blueprints exist and are important is made much clearer through the discovery of the genetic chains as codes. Nobody in the 19th century or even now would profess to understand the details of how, from the code, an actual organism is produced.
>
> SCHÜTZENBERGER: We are not worried with the details. The only thing is that I would need an example where such a correspondence would exist or could exist, even in the simpler case.   *(Moorhead and Kaplan 1967, 75)*

This ambiguity aside, I think we can reconstruct Schützenberger's argument like this: Evolutionary theory is fundamentally flawed because it has no account for how random changes to the genotype translate into changes in the phenotype, or even how it is possible for such a process to work. Moreover, when we try to simulate this process on a computer, we find that it fails completely. Random changes to computer programs not only do not lead to improved programs, they nearly always lead to nothing functional at all.

The other attendees were entirely unimpressed by this argument, for good reason. It is important to distinguish two different questions. One question is this: What are the physical mechanisms through which the sequence of genetic letters in a genotype lead to the creation of an actual organism? Everyone at the conference agreed that those mechanisms were mysterious, and also that those mechanisms were relevant to understanding evolution. Had Schützenberger stopped here the discussion following his talk would have been far more harmonious.

Instead he acted as though he was addressing a different question: Does our current ignorance of the mechanisms of development imply that evolutionary theory is just fundamentally unsound? The answer to *that* question is no. Ulam made this point well in the discussion:

> What you are saying, it seems to me, is that the Darwinian and
> neo-Darwinian theories are not complete, and everybody agrees
> with that; but it is not an objection to the scheme of things, which
> is sort of lost sight of.   *(Moorhead and Kaplan 1967, 76)*

Ulam's point can be understood like this: It is often said that in science there are problems and there are mysteries. "Problems" are open questions that are likely to be resolved through more research of the sort that is already ongoing. "Mysteries" are open questions that seem utterly incomprehensible within the current best understanding of science. In saying that Schützenberger's arguments were not an objection "to the scheme of things," Ulam was saying the then current ignorance of embryological development was merely a problem and not a mystery.

Biologist Richard Lewontin also made some salient points during the discussion. As part of an exchange about how random mutations affect phenotypes, he said:

> [I] can give [Schützenberger] known cases where the enzyme, far
> from being destroyed, is changed in its pH optimum, changed in
> its isoelectric points, changed in a number of aspects of its
> physiological function by single substitutions of single amino

> acids. We know exactly where in the phenotypic topology of the protein these amino acids have been substituted, and we can specify exactly in what way they change the physiology of the organism, changing its fitness in the write-in space.
>
> *(Moorhead and Kaplan 1967, 76)*

In other words, what Schützenberger described as fundamentally inconceivable was actually just a simple empirical fact. Random mutations in genes really do lead to observable changes in organisms, and this was well-known even in 1966. To the extent that we are discussing the in-principle soundness of the theory, this is all that matters.

Let us move now to the analogy with computer programs. If we make random changes to the code, the result is nearly always worthless gibberish. This is because the words of the language in which the code is written are inevitably highly isolated from one another in the space of all possible words. This point is more easily seen in the context of a natural language like English. If we randomly change one letter in a given four-letter word, the result is very likely to be something that is not an English word. It turns out there are roughly 4,000 four-letter words in English, but there are more than 450,000 possible sequences of four letters. Moreover, the tiny fraction of meaningful words is scattered more or less at random throughout the space of all possibilities. (There are certainly some regularities among four-letter English words, but for our purposes we can treat the arrangement as random.) If the meaningful words were grouped very close together, then we might hope that a random change to one letter might bring us to a different word, but that seems not to be the case, at least not in general.

From these examples, we learn something interesting about English and certain computer languages, but we are not learning general lessons that have to apply to all languages. To the extent that the genetic code can be viewed as a language at all, it is a very small one. The words of the language are triplets of codons, each of which can take on one of four values. That makes a grand total of 64 words, and every one of them is meaningful within the language. This point was made by several of the conference participants. For example, Lewontin said to Schützenberger:

> I think the answer is that you have over-estimated the number of absolutely meaningless changes that occur when you change a single nucleotide. If we list all single nucleotide changes and the known translation vocabulary between nucleotide triples and insertion of amino acids, and then we list for a given protein all the results on that protein of changing amino acids all over the molecule, we will find, in fact, that a very large proportion of those do not render the molecule meaningless in an absolute context.   *(Moorhead and Kaplan 1967, 79)*

Conrad Waddington later made a similar point:

> Before we go any further, I think that, first of all, we should agree how we are using the word "meaningful." I think Schützenberger means that when he changes something in program space, nothing comes out at all. . . . But actually when we change something, *some* protein does come out; it may not be a very good protein, but some protein comes out. All proteins do something, so all changes in the program level have meaning, in the sense that they produce a protein, except for some full stop marks and so on.
> *(Moorhead and Kaplan 1967, 79)*

This seems like a salient point. In contrast to Schützenberger's computer programs, the genetic machinery does not always, or even usually, jam in response to random changes. The genetic language is far more resilient to change than Schützenberger's computer languages, and therefore they are not analogous in the relevant way.

We can dramatize this point with an analogy. In Section 1.3, I mentioned having lived in the state of Kansas for several years.

Kansas has one of the lowest population densities of any US state. When I first moved there, I was surprised to discover that once you left one of the state's major population centers, it might be many miles in all directions before you hit another major town. I had grown up in the state of New Jersey, which is actually the most densely populated state in the country. Moreover, I spent most of my time in the central and northern parts of the state, which are especially densely populated areas. If you are driving in this portion of New Jersey and you leave a town, then most of the time you are immediately in another town. If there is no sign on the road to tell you otherwise, you will not know that you have left one town and entered another.

Kansas is like Schützenberger's computer programs, and Schützenberger himself is like someone who has only visited Kansas and not New Jersey.

Physicist Victor Weisskopf perfectly summarized the discussion, showing some impatience with Schützenberger:

> I want to analyze the difference of opinion between Schützenberger and the rest of the world. This is, I think, the following: Schützenberger says that in the typographical space, the overwhelming number of changes that can be done at random have absolutely no meaning, and he puts in support of it the fact that if you have a computer, and you change the program at random, it always is destroyed.
>
> The other side says that that isn't so. The kind of program which genetics has produced with the 3-letter code is such that it isn't so. I think this is what Lewontin says, that a lot of changes, maybe not an overwhelming number but a large percentage, do make sense in the biochemical sense of the word, and here I think is the discrepancy.   (Moorhead and Kaplan 1967, 80)

It would seem, therefore, that Schützenberger's argument is based in large measure on a bad analogy, and it can be dismissed on that basis.

4.4   THE PERILS OF LONG-TERM MODELING

Modern anti-evolutionists periodically rediscover the Wistar conference. In their telling, the conference involved heroic physicists and
mathematicians explaining how the world works to ignorant and
dogmatic biologists. Representative is the following description from
ID proponent and law professor Phillip Johnson. Referring to mathematical attempts to determine evolution's fundamental soundness,
he writes:

> Some mathematicians did try to make the calculations, and the
> result was a rather acrimonious confrontation between themselves
> and some of the leading Darwinists at the Wistar Institute in
> Philadelphia in 1967. ... For example, the mathematician D. S.
> Ulam argued that it was highly improbable that the eye could
> have evolved by the accumulation of small mutations, because
> the number of mutations would have to be so large and the time
> available was not nearly long enough for them to appear. Sir Peter
> Medawar and C. H. Waddington responded that Ulam was doing
> his science backwards; the fact was that the eye *had* evolved and
> therefore the mathematical difficulties must be only apparent.
> Ernst Mayr observed that Ulam's calculations were based on
> assumptions that might be unfounded, and concluded that,
> "Somehow or other by adjusting these figures we will come out all
> right. We are comforted by the fact that evolution has occurred."
>
> The Darwinists were trying to be reasonable, but it was as if Ulam
> had presented equations proving that gravity is too weak a force to
> prevent us all from floating off into space. Darwinism to them
> was not a theory open to refutation but a fact to be accounted for,
> at least until the mathematicians could produce an acceptable
> alternative.   *(Johnson 1991, 38–39)*

The Wistar conference was held in 1966, not 1967 (the proceedings
were published in 1967), and Dr. Ulam's initials were S. M. (for
Stanislaw Marcin) and not D. S., but these are hardly the most

serious problems with this statement. Johnson's description is not quite factually wrong, but it is so misleading that it amounts to the same thing. Anyone reading this would think Ulam presented a strong, mathematically precise, argument for the insufficiency of evolutionary theory, but was rebuffed by dogmatic and uncomprehending Darwinists. In reality, that is not at all what happened.

Ulam's presentation bore the slightly awkward title, "How to Formulate Mathematically Problems of Rate of Evolution?" In contrast with Eden and Schützenberger, Ulam was not challenging the fundamental soundness of the theory. As expressed in his talk, his attitude toward neo-Darwinism is best described as agnostic. His intent was to investigate the question of whether there was sufficient time in natural history for Darwinian evolution to accomplish what was attributed to it, and he wanted to develop a mathematical model useful for that purpose. The modesty of his intentions is made clear by this statement, from the beginning of his talk:

> [I] have done a bit of very schematic thinking on the mathematics of such a process, and I want to make some remarks to you which are not, as one of the speakers stressed before, correct in a realistic sense, but might be relevant for the approach to some quasimathematical discussion at least. The philosophical and general methodological remarks made by various speakers so far can form a basis of what can be, sometime in the future, mathematized. What I am going to do will consist, as it were, of picking out various items from the comments made so far and try to show how, perhaps in some remote future, mathematical schemata can be formulated. *(Moorhead and Kaplan 1967, 21–22)*

This sort of intellectual modesty is ubiquitous throughout the presentation. For example, after stating that he will discuss certain variables and parameters that are relevant to modeling evolution mathematically, Ulam says:

> In my talk, I will give you a whole set of such parameters with
> values it is important to know. The trouble is that at present
> realistic definitions of these parameters, not to mention the
> numerical values, are completely unknown.
>
> *(Moorhead and Kaplan 1967, 22)*

These are not the remarks of someone who believes he has a mathematical refutation of evolution.

He went on to identify some of the important variables that
govern the rate of evolution. For example, we need to know the sizes
of the populations with which we are dealing, the average life span of
the members of the population, the amount of time available for the
whole process, the frequency of favorable mutations, and the selective
advantage a given favorable mutation might give to its bearer. When
we have in mind a specific biological system, we must also estimate
the number of genetic changes required to go from an ancestral state
lacking the system to a descendant state possessing it.

Ulam used the eye as his model system, and he suggested
possible values for some of the relevant parameters. With respect to
this part of the presentation, there were two items in particular that
drew objections from the biologists, especially Mayr.

The first was Ulam's suggestion that if the eye had evolved
through a large number of small improvements, then each individual
improvement would have only a very small selective advantage. Mayr
objected to this, citing then recent experimental work showing that
it was likely the selective advantages would have been much higher
(thereby speeding up the evolutionary process considerably).

The second was when Ulam suggested that the numerous small
improvements leading to the evolution of the eye had to happen in
succession. His model entailed that one improvement had to appear
and spread through the population before the next one could arise.
Mayr objected to this as well, suggesting that to a large degree they
could all go on simultaneously. There was some back and forth at this

point in which Ulam suggested such a scenario was unlikely and Mayr disagreed, but the discussion quickly went off in a different direction, and the issue was not explored.

As it happens, precisely this issue arose in a portion of Eden's presentation that I did not discuss in Section 4.2. Specifically, at one point Eden suggested that the improvements on which evolution relied had to come in succession, with one becoming fixed in the population before the next one could arise and spread. Waddington, in a summary paper delivered at the end of the conference, gave the correct response to this. His statement applies equally well to Ulam's remarks and shows that Mayr was right to object:

> The point was made that to account for some evolutionary changes in hemoglobin, one requires about 120 amino acid substitutions. The calculations were done on the basis of treating these substitutions as individual events, as though it is necessary to get one of them done and spread through the population before you could start processing the next one. The whole thing has to be done in sequence; and of course, if you add up the time for all those sequential steps, it amounts to quite a long time.

> But the point the biologists want to make is that that isn't really what is going on at all. We don't need 120 changes one right after the other. ... Calculations about the length of time of evolutionary steps have to take into account the fact that we are dealing with gene pools, with a great deal of genetic variability, present simultaneously. To deal with them as sequences of individual steps is going to give you estimates that are wildly out.
>
> (Moorhead and Kaplan 1967, 96)

We now move on to the statements made by Waddington and Medawar. Johnson would have you believe they responded to Ulam's insightful mathematics by refusing even to consider it, demanding that evolution be taken as an unassailable starting point. In reality, their remarks came in the discussion after Ulam's presentation, and after the criticisms of his model just described had been raised. It was in this context that Waddington said:

> Could I put your question upside down? You are asking, is there enough time for evolution to produce such complicated things as the eye? Let me put it the other way around: Evolution has produced such complicated things as the eye; can we deduce from this anything about the system by which it has been produced? One possible deduction would be that this thing worked by algorithms rather than by describing bits. ...
>
> *(Moorhead and Kaplan 1967, 28)*

This certainly does not sound like an expression of mindless dogmatism. It sounds more like Waddington is suggesting a potentially fruitful starting point for an investigation, one more likely to be successful than what Ulam put forth. Waddington and Medawar were simply putting into practice the principle we discussed in Section 1.3: If extensive physical evidence suggests that something occurred, but a simplistic mathematical model says it is impossible, then it is the model and not the evidence that must yield.

Which brings us, finally, to Mayr's statement. No one familiar with anti-evolution literature will be surprised that Johnson has omitted crucial context for understanding Mayr's point. Johnson even shortened a sentence without giving any indication that he had done so. Mayr's full statement was this:

> So all I am saying is we have so much variation in all of these things that somehow or other by adjusting these figures we will come out all right. We are comforted by knowing that evolution has occurred. *(Moorhead and Kaplan 1967, 30)*

Had Johnson quoted the full sentence, his readers might have wondered what Mayr meant by "we have so much variation in all of these things." And that, in turn, would have required providing some further context.

Here is that context: The statement I just quoted comes at the end of a very long speech by Mayr in the discussion after Ulam's presentation. Mayr spoke for several minutes, uninterrupted. In this speech, he went through several of Ulam's parameters and pointed to concrete reasons why it was so hard to assign numerical values to them. The parameters themselves, he argued, are influenced by so many other variables that they can take on wildly different values in different contexts. His point was twofold. The first was that Ulam's model was just too simplistic to be useful for anything. The second was that, precisely because the relevant parameters are so numerous and variable, mathematics is not useful for answering a question such as, "Did evolution have enough time to produce an eye?"

Summarizing, here is what actually happened at Wistar: Ulam presented a mathematical model for studying rates of evolution that, by his own repeated admission, was simplistic, not biologically realistic, and intended only as a starting point for discussion. The biologists then pointed out specific, concrete ways in which the model was unrealistic. They were able to do this so readily because of their familiarity with recent experimental work in this area and their immersion in the relevant theoretical concepts. They then suggested a better approach to devising a mathematical model for something like the evolution of the eye.

This gives rather a different impression than that gained from Johnson's account.

I have belabored this discussion for two reasons. The first is that it is a specific example of a general phenomenon in anti-evolutionist literature: a complete lack of conscience about accurately presenting the views of their opponents. Johnson, you can be sure, had no interest in the technical minutiae of what was discussed at Wistar. Instead he just selected a few out-of-context phrases he could use for rhetorical effect and presented a highly distorted view of what happened, strictly for the purpose of making biologists look bad. This sort of thing is entirely standard among anti-evolutionists, which is one of the reasons scientists generally view them with such disfavor.

The second reason is that it perfectly illustrates one of the themes of this book: distinguishing good mathematical modeling from bad mathematical modeling. There is a reason biologists do not engage in the sort of long-term modeling attempted by Eden and Ulam.

Earlier, we said that mathematical modeling is the art of discarding most of reality, in the hope that what little remains is the really important part. Particularly in physics, this sort of thing is often possible. When seeking useful solutions to physical problems, it is often true that you only need to consider a small number of variables. In contrast, when you are studying the possible outcomes of millions of years of evolution, you need to keep all the variables. There is no useful mathematical model to be found because all the myriad variables are important.

This was precisely Mayr's point. He certainly was not saying that Darwinism is adhered to dogmatically as an unquestioned axiom. Instead he was saying something more like this: We have copious physical evidence that modern life forms are the end results of a lengthy evolutionary process. Since this implies the eye is the product of evolution along with everything else, we accept that as a working hypothesis. And since no mathematical model could possibly include enough of nature's complexity to be convincing, we are not worried that a few back-of-the-envelope calculations will provide a good reason for abandoning that hypothesis.

4.5 THE TWO PILLARS OF MATHEMATICAL ANTI-EVOLUTIONISM

The proceedings of the Wistar conference are sometimes fascinating and sometimes frustrating. There are moments of great lucidity and insight, but also moments where you feel people are talking past each other or failing to make important points. This is to be expected when reading transcripts of in-the-moment conversations, as opposed to polished papers.

For me, the conference heroes were Ernst Mayr and Conrad Waddington. Mayr's own presentation bore the witty title, "Evolutionary Challenges to the Mathematical Interpretation of Evolution." It was a brilliant elaboration on the theme of what goes wrong in trying to devise long-term mathematical models for evolution. Waddington supplied an eloquent summary paper that presented the issues with a lucidity that had sometimes been lost during the formal presentations.

They both participated in many of the discussions, and most of what they said was cogent and on point. For example, I had to smile when I came across this statement from Waddington's summary paper at the end of the conference. Bear with me, since this is worth quoting in detail:

> I think what the biologist is saying in this connection is that we have a space of all possible nucleotide sequences with associated amino acid sequences, so that you have the DNA and the protein which you can consider as a complement. ... This set was entered at the beginning of life, starting at some point or points and was explored to find sets which would operate adequately enough to ensure transmission. The space was explored to some extent, but always sequentially from the last position to some neighboring position.
>
> As soon as it was explored enough to get sufficient of these couples together to work with fair efficiency, they became insulated from the whole of the rest of the space ... Therefore, life has only explored a minute fraction of the total nucleotide space.
>
> In the part which it has explored, which is the only part relevant to evolution, biologists are asserting that the meaningful section of it is quite large in comparison with all the things that could conceivably be made out of it in single steps. It is a much larger fraction than is the space of meaningful strings of English words. I think this is the point. We are asserting that it is a large fraction

of the total space which could be made from the nucleotides involved, but still we are saying that the meaningful space is a minute fraction of the total nucleotide space.

(Moorhead and Kaplan, 93)

That is exactly right and perfectly summarizes why the biologists were so unimpressed by what Eden and Schützenberger had to say.

The modern legacy of the Wistar conference has three parts. The first is that it brought to prominence the strategy of modeling evolution as a combinatorial search so that mathematics can be deployed against it. As we have noted, this is the strategy underlying all of the major strands of mathematical anti-evolutionism today. The second is that we see the two main strategies that are used to implement the model. We can either follow Eden by carrying out a calculation of some sort, or we can follow Schützenberger in looking for general principles that will militate against evolution's plausibility. The final point is that we see in their presentations the general errors that have proven fatal to all such arguments proposed to date. Those following Eden have been unable to formulate a biologically reasonable calculation, while those following Schützenberger run afoul of empirical considerations.

One of the participants at the Wistar conference was botanist J. L. Crosby. Exasperated by what he felt were manifest errors in Ulam's presentation, he unleashed a broadside that will serve as a fitting coda to this chapter:

It always seems to me that there is such an air of gorgeous unreality when mathematicians come to deal with biological subjects and I think that is the case here. ...

I feel that mathematicians often take ideas of biology to amuse themselves, rather than to advance knowledge. They are quite welcome to take our ideas, which must present them with an enormous field of possibilities for mathematical maneuvering; but I do really think that if they want us to take these things

> seriously, they have to present them in a way in which not only
> do we understand them but in which they make biological sense.
>
> *(Moorhead and Kaplan 1967, 31)*

4.6   NOTES AND FURTHER READING

The Wistar conference had no significant impact on the work of
biologists, and it received almost no scholarly attention at all after it
ended. A perfunctory review of the conference proceedings appeared
in the journal *Science* in April 1968, written by botanist John L.
Harper. He wrote:

> Most biologists are satisfied with a theory that can be tested and
> that proves predictive. It is a different challenge to a theory that it
> should have an effective working model, for failure may imply
> either imperfection in the theory or imperfection in the model. It is
> doubtful whether this symposium has done much to influence the
> theory of evolution; it may have done much to improve future
> models.   *(Harper 1968, 408)*

The first part of Harper's conclusion has proven true, but the second
part has not.

Notwithstanding the manifest weakness of Eden's and Schützenberger's
arguments, anti-evolutionists frequently tout the Wistar conference in
their writing. A recent example is ID proponent Stephen Meyer, who
devotes nearly eight pages to the conference in his book *Darwin's
Doubt* (Meyer 2013). We shall consider some of what he said about the
conference in Chapter 5.

Molecular biology and computer science were both in a very rudimentary
state when the conference was held in 1966. Both fields have
blossomed spectacularly in the ensuing decades, with results that
have not been kind to the perspectives of Eden and Schützenberger.
Molecular biologists have learned a lot about the geometrical
structure of protein space, and their results make Eden's simplistic
combinatorial calculations look hopelessly naive. Schützenberger's
arguments about computer simulations have likewise not held up,
and computer simulations of the evolutionary process are now
commonplace. We will discuss this work in Chapter 6.

In previous writing, I provided documentation regarding the propensity of anti-evolutionists to engage in extremely dishonest rhetorical strategies, such as quoting scientists out of context to make them appear to be saying something entirely different from their actual point (Rosenhouse 2002a, 2005). The webpage edited by John Pieret (2006) is also very useful for this purpose. It seems that this has long been an unsavory aspect of anti-evolutionist writing. In a 1973 essay, geneticist Theodosius Dobzhansky wrote:

> Disagreements and clashes of opinion are rife among biologists, as they should be in a living and growing science. Antievolutionists mistake, or pretend to mistake, these disagreements as indications of dubiousness of the entire doctrine of evolution. Their favorite sport is stringing together quotations, carefully and sometimes expertly taken out of context, to show that nothing is really established or agreed upon among evolutionists. Some of my colleagues and myself have been amused an amazed to read ourselves quoted in a way showing that we are really antievolutionists under the skin. *(Dobzhansky 1973, 129)*

A few years after the Wistar conference, Ulam published a lengthy paper outlining his ideas for mathematizing various biological problems (Ulam 1972). In those sections of the paper that address evolution, Ulam expresses no skepticism about it at all. Instead he writes as though he regards phylogenetic reconstruction (working out the precise evolutionary relationships among both modern and fossil species) as an integral and legitimate aspect of biological practice.

As suggested by his presentation at the Wistar conference, Ernst Mayr was generally skeptical of mathematical treatments of evolution. In particular, he derided the mathematical models of population genetics as "beanbag genetics," and believed that such models were too simplistic to be useful for anything. While Mayr was a giant of twentieth century biology, on this point he was in the minority among his colleagues. One of his main foils in this debate was J. B. S. Haldane, who pioneered the mathematical treatment of evolution. For a readable account of their points of view, try the book by Dronamraju (2011).

5   Probability Theory

We have noted that evolution seems implausible to many because it runs afoul of our intuition that natural forces are unlikely to build complex, functional structures, no matter how much time you give them. Mathematical anti-evolutionism is largely an attempt to provide a rigorous foundation for that intuition. Since probability theory is the branch of mathematics that quantifies how likely or unlikely we believe an event to be, it naturally plays a big role in anti-evolutionist literature.

Probability can be a tricky subject, so let us try a warm-up exercise. Suppose you have two well-shuffled decks of cards, and you turn over the top card on each deck. What is the probability that you get the ace of spades on at least one of the decks? Think about that for a moment before reading on.

Here is a plausible argument: Each deck has 52 cards, only one of which is the ace of spades. That means the probability of turning over the ace of spades on each deck individually is $1/52$. And since the two decks are independent of one another, we can add these probabilities to get

$$\frac{1}{52} + \frac{1}{52} = \frac{2}{52} = \frac{1}{26}.$$

What do you think?

Of course, if that really were the answer, then the problem would not be worth mentioning. The problem is interesting precisely because that seemingly impeccable argument overlooks an important point, illustrated in Figure 5.1. Informally, the problem is that we do not win as often as we think we do because sometimes we get the ace

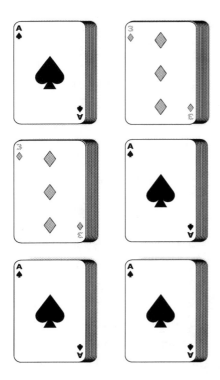

FIGURE 5.1 If the ace appears on the left deck with a random card on the right (the three of diamonds in this example), then we win. And if the ace appears on the right with a random card on the left, we also win. But if the ace appears on both decks simultaneously, then we have wasted one of our rare and precious wins.

of spades on both decks. When that happens, it is as though we wasted one of our rare and precious wins. We win so long as either top card is the ace of spades, but we do not get a double win if both top cards are aces.

We can quantify this. Since there are 52 possible top cards for the first deck, and 52 possible top cards for the second deck, we see that there are $52 \times 52 = 2,704$ possible pairs that we might get when we flip over the cards.

How many of those pairs have an ace of spades in them? Well, we can get the ace of spades on the first deck paired with any card on the second, and that makes 52 pairs. Likewise, we can get the ace of spades on the second deck paired with any card on the first, and that makes another 52 pairs, for a total of 104. However, the pair in which both cards are the ace of spades got counted twice, meaning we need to subtract 1 to get 103.

So the answer is $103/2{,}704$, which is very slightly less than $1/26$. (If you care to check, $104/2{,}704$ is exactly $1/26$.)

That was just a brainteaser, but it raises an important point. Probability can be surprisingly subtle. In particular, when you carry out a probability calculation, you need to be very precise about the space in which you are working.

As we shall see, anti-evolutionist probability never heeds this warning.

5.2   A PROBABILITY PRIMER

Recall that in Section 3.2, we made a distinction between track one and track two mathematics. Track one is about having an intuitive and general understanding of the mathematical concepts. Track two is about expressing everything with care and precision, and this usually necessitates the use of mathematical notation. We emphasized that both tracks are important when assessing a piece of mathematics. Without track one it is difficult to understand what the symbols are saying, and without track two the conversation is inevitably too vague and imprecise for scientific purposes.

Here, and in the next two chapters, we turn to the arguments at the heart of modern mathematical anti-evolutionism. A proper assessment of these arguments requires getting our hands dirty with some technical details. In most cases, a track one discussion will be sufficient to expose the anti-evolutionist's conceptual errors. In certain cases we will refer to some track two details, but it will always be possible to follow the flow of the argument even if these details become too dense.

With that said, let us return to our discussion of probability. A track one approach to the subject begins with the observation that sometimes long runs of events are predictable even though individual events are not. For example, we may not know if a fair coin will land heads or tails, but we know that in a large number of tosses it will land heads roughly half the time and tails roughly half the time. Likewise, if we draw a card from a well-shuffled deck, then we do not know what suit we will get. But we *do* know that if we choose cards over and over again, then each of the four suits will appear roughly one fourth of the time.

Moreover, these conclusions are intimately related to counting. The coin toss has two possible outcomes – heads or tails – and each is equally likely. That is why we get heads half the time and tails half the time. In our second example, we know that each of the four suits is represented 13 times among the 52 cards. We take that to mean that each suit will be chosen roughly 13 out of every 52 times, which is the same as one time out of every four.

A slightly more complicated example involves rolling two six-sided dice. If we roll the dice, then what is the most likely sum? Naively, we might argue that there are 11 possible outcomes since the sum can be anything from 2 to 12. If these sums are treated as equally likely, then we might think each sum occurs with probability 1 out of 11. However, it is not hard to spot the error in this reasoning. There are actually 36 possible outcomes, not 11, since each of the two dice has six possible outcomes on its own. Among those 36 possible outcomes, we can quickly count that there are six ways to get a sum of seven. (We could get 1 on the first die and 6 on the second, 2 on the first die and 5 on the second, and so on.) When we do a similar count for all the other possible outcomes, we find that none occurs more often than a sum of 7.

Let us push this a little further. What happens if we want the probability that two events will happen together? For example, suppose that I am tossing a coin on one side of the room while you are rolling a die on the other. What is the probability that I toss a head at the same time that you roll a 3? Our reduction of probability to counting handles this case well. There are twelve possible outcomes for this experiment: I can get a head paired with any of the six numbers on the die, or I can get a tail paired with any of the six numbers on the die. These twelve outcomes are equally likely, so the probability of getting a head with a 3 is 1/12.

The interesting part is that I could have arrived at the same conclusion by multiplying the individual probabilities. The probability of getting heads is 1/2, and the probability of rolling a 3 is 1/6. When I multiply these together I get 1/12, same as before. Note that there are two possible outcomes for tossing a coin and six possible outcomes for rolling a die, and that makes twelve possible outcomes for the two activities taken together.

We can even do an abstract proof of this. Suppose that one event happens with probability p/q and a separate event occurs with probability r/s. Then we can take this to mean that there are q possible outcomes for the first experiment and s possible outcomes for the second experiment, which makes qs possibilities for the two experiments taken together. Likewise for the top of the fraction: the first event can happen in p possible ways, while the second event can happen in r possible ways, and this makes pr possible ways for the two events to happen together.

This is a very useful principle: If two events are independent, then the probability that both events occur is found by multiplying the individual probabilities. But this only works when the events are independent! If the events are related to one another, then multiplying the probabilities gives the wrong answer.

For example, suppose we draw cards from a deck and we want the probability we choose a card that is both red and a heart. We certainly could not multiply the probabilities together since being a heart forces you to be red. The two events are related to one another, and therefore they are not independent. Indeed, the probability of drawing a red card is 1/2, and the probability of drawing a heart is 1/4. Multiplying these together gives 1/8. But the correct answer is 1/4, since the probability of drawing a red heart is just exactly the same as the probability of drawing a heart, and since exactly one quarter of the cards in a deck are hearts.

This is an important point since, as we shall see, anti-evolutionists sometimes multiply probabilities together without first verifying that the underlying events are independent.

These examples are already sufficient for a track one treatment of probability. Our intuition is that long runs of identical trials might be broadly predictable even when individual outcomes are not. We can often get at these long run patterns by counting up possibilities among equally likely outcomes. But we have to be careful about how we do our counting since there is a danger of treating outcomes as equally likely when really they are not.

This shows us what is needed as we pass to a track two treatment of probability. A proper probability calculation must begin with a rigorous accounting of all possible outcomes, coupled with an initial assignment of probabilities to each outcome. The manner in which we assign the probabilities is referred to as the "probability distribution" for the outcomes. When we have a properly enumerated list of outcomes coupled with an appropriate distribution, we say that we have a well-defined "probability space." In other words, we cannot talk seriously about probability until we have defined the space in which we will work.

The concept of a distribution will be especially important going forward, so let us pause to consider it more carefully. When we assign the same probability to every outcome, we say that we have used the "uniform distribution." For example, we are using the uniform distribution when we assign heads and tails a probability of $1/2$ when we flip a coin. We use it again when we assign a probability of $1/6$ to each of the six numbers on a standard die.

On the other hand, the uniform distribution is not always appropriate. When we roll two dice, we can take our space of outcomes to be the eleven numbers from 2 to 12. We have already seen that of the 36 ways two dice can land, six of them correspond to a sum of 7. We can quickly see that 2 and 12 can only happen one way each (snake eyes or boxcars). Totals of 3 and 11 can happen two ways each (1, 2 and 2, 1 for a 3, and 5, 6 and 6, 5 for an 11). Continuing in this way, we see there are three ways of getting 4 or 10, four ways of getting 5 or 9, and five ways of getting 6 or 8. Putting everything together, we get the probability distribution shown in Table 5.1. If you are unclear

Table 5.1 *The probability distribution for the possible sums when rolling two dice. Note that the first row should be taken to mean that P(2) = P(12) = 1/36, the second row means that P(3) = P(11) = 1/18, and so forth.*

Outcome	Probability
2 or 12	$\frac{1}{36}$
3 or 11	$\frac{1}{18}$
4 or 10	$\frac{1}{12}$
5 or 9	$\frac{1}{9}$
6 or 8	$\frac{5}{36}$
7	$\frac{1}{6}$

on how we arrived at some of these fractions, note that $2/36 = 1/18$, $3/36 = 1/12$, and so on.

In simple examples involving coins, playing cards, or dice, choosing the right distribution seems pretty straightforward. However, choosing the right distribution can be a very difficult practical problem when dealing with data sets drawn from complex, real-life situations. If you decide to pursue a career in statistics, much of your training will involve learning how to discern the probability distribution appropriate to different kinds of data. For us, however, the point is this: A serious probability calculation can only be carried out within a properly defined probability space, and the burden is on the one doing the calculation to justify their choice of distribution.

Now suppose you are an anti-evolutionist. You start with a track one intuition that there is *something* improbable in the idea of evolution, or any other naturalistic process, crafting complex, functional structures. But you also know that scientists will not take seriously a mere statement of incredulity. So you go looking for a track two argument. This entails defining a proper probability space, which in turn entails enumerating the outcomes and assigning probabilities to each.

Anti-evolutionists have tried a number of approaches to meeting this burden, but we shall argue that none are at all successful. However, we must attend to some further preliminaries before examining their arguments.

5.3   THE HARDY—WEINBERG LAW

Most of this chapter is devoted to the poor probability arguments put forth by anti-evolutionists. As preparation for this, let us first consider a reasonable application of probability to evolution.

Evolutionary biology addresses fundamental questions about the origins of humanity, and that is one of the reasons it receives so much attention from nonscientists. However, the daily work of most biologists is not about anything quite so grand as ultimate origins. Much of the routine work in evolution is devoted to mundane questions of the following sort: Given a population of organisms and some knowledge about the frequencies of various genes, what can we say about the gene frequencies a few generations hence? Since such questions involve the genetics of populations, the branch of evolutionary biology that studies them is known as "population genetics."

How might we develop a mathematical model to study this question?

A typical scenario in genetics is for an animal's chromosomes to contain two copies of a gene that can exist in two different forms. If we refer to these genes as *A* and *B*, then we can say that each animal is either of type *AA*, *BB*, or *AB*, depending on whether the two copies of the gene in that animal are the same or different. We can then imagine listing the types for all the animals in the local population, which would produce a big collection of *A*s and *B*s. From this collection, we could work out the percentage of the total that are *A*s, as well as the percentage that are *B*s.

For example, if there were five animals in the population of types

$$AA \quad AA \quad AA \quad AB \quad BB,$$

then 7 of the 10 genes are As and 3 of the 10 are Bs. That makes 70% As and 30% Bs.

It is customary to refer to these percentages as "frequencies" and to represent them as decimals. In my example, the frequency of A is 0.7 and the frequency of B is 0.3. By writing things this way, we have the convenient fact that the frequencies add up to 1. We can also think of frequencies as probabilities, in the sense that if we reach into our bag of As and Bs and pull one out at random, the probability of pulling an A is 0.7, and the probability of pulling a B is 0.3.

Return now to the general set up. We have the three types, AA, AB, and BB, and we will assume that A and B appear with frequencies p and q, respectively. (Of course, we know that $q = 1 - p$, but it will be simpler just to use the single letter q rather than the more complex $1 - p$.) If we now let the animals do what they do, how will the three types be represented in the next generation?

Our answer to this question will depend on the answers to several other questions:

- Do the animals choose their mates at random with respect to A and B, or do they prefer to mate only with others of the same type?
- Do all three types have the same chance of finding a mate, or does one of the types have an advantage over the others?
- Do we want to allow for the possibility that this gene will mutate in the next generation? Are we allowing for possible migration into and out of the population? These factors would change the gene frequencies in the next generation.
- Can we assume the population is very big, so that there is a large number of couplings in each generation? If we do not make this assumption, then there is a danger that the gene frequencies will change in the next generation just by chance.

If that last point is unclear, think about flipping a coin. It would not be terribly surprising if we flip the coin three or four times and get all heads or all tails. In short runs, the frequencies of heads and tails might differ substantially from fifty-fifty. But in a long run these frequencies will eventually settle down to their expected values. In the biological context, in a small population, gene frequencies might change just by luck in the same way that we might toss three heads in a row. But in a large population this is much less of a factor.

At this point we should recall what we said about mathematical modeling: we throw out most of the messy reality and then hope the bit that remains includes the really important stuff. So, let us assume that the animals do, indeed, choose their mates randomly, that they all have an equal chance of finding a mate, that there is no mutation or migration, and that the population is very large.

What are the consequences of these assumptions? Each of the offspring in the next generation will inherit one copy of the gene from its mother and one from its father. Since each gene individually has a probability p of being A, and since the gene inherited from the mother is independent of the gene inherited from the father, we can multiply these together to get a probability of p^2 that the offspring is of type AA. By similar reasoning, the probability is q^2 that the offspring is of type BB.

Working out the probability of getting type AB is a little trickier because there are two ways for this to happen. It might be that the offspring inherits A from the mother and B from the father. This will happen with probability pq. But it might happen instead that the offspring gets B from the mother and A from the father, and this also happens with probability pq. Adding these together gives us $2pq$, and that is the probability that the offspring is of type AB.

Our model predicts that a randomly chosen offspring in the next generation will be type AA with probability p^2, type BB with probability q^2, and type AB with probability $2pq$. But notice that this conclusion depended only on the values of p and q and did not depend on the proportions of the three types in the initial population. And since the values of p and q have not changed from the parent's to the child's generation, because no As or Bs were gained or lost in the transition, the representation of the three types will remain constant for all subsequent generations.

(For the sake of accuracy, let me mention that I have omitted a few technical details that very slightly affect this conclusion, but including these details would be more trouble than it is worth relative to my purposes in this section. I say a little more about this in Section 5.11.)

This finding – that the three types will be represented with probabilities p^2, q^2, and $2pq$ in the next generation – is known as the Hardy–Weinberg law, after the mathematician (Hardy) and geneticist (Weinberg) who first published it. The law's conclusion was based on such a simplistic model that you might think it could not possibly be useful for anything. But you would be wrong! There have been many studies of gene frequencies in natural populations, and with surprising regularity the findings are in accord with what the Hardy–Weinberg law tells us to expect. When this happens, we conclude that the assumptions underlying the model hold true in the population with regard to that gene. When the data is not in accord with the law, we conclude that one of the assumptions does not hold. Either way, we have learned something about the animals we are studying.

We could make our model more realistic by introducing some further variables. In real populations there is immigration and emigration, and this affects the frequencies of A and B. These frequencies could also change as a result of mutation. Most importantly, it might be that there is a selective advantage to possessing either A or B, and when this is the case the Hardy–Weinberg frequencies certainly will not hold in the next generation. We could construct more elaborate models that take these variables into account, and population geneticists have been very assiduous at doing precisely that. Alas, since each of these variables is inevitably represented by a different letter in the ensuing equations, textbooks on population genetics quickly come to look like alphabet soup. The mathematical formalism is formidable, but for our purposes it will not be necessary to consider the details.

Models of this sort are useful to biologists engaged in field work because they provide a starting point for research into the genetic makeup and behavior of actual populations. They also have theoretical value, in that they can help reveal what is possible under different assumptions. For example, in the early twentieth century, many scientists questioned whether natural selection was a powerful enough force to account for large-scale evolutionary change. If a genetic variation gave one animal a tiny reproductive advantage over another, would selection really be able to cause its spread through the population? Mathematical models were able to resolve this question by showing that under a variety of plausible assumptions, even minuscule selective advantages would be sufficient to drive the gene to fixation (meaning, essentially, that every animal in the population comes to possess that gene).

We should also be mindful of the domain of application for these models. Over a small number of generations, the variables represented in the model are likely to remain constant, and it is for this reason that population genetics is sometimes said to model short-term gene flow. However, as time passes environments and populations change, to the point where the models are no longer useful. That is why these models are not used for predicting the long-term course of evolution.

Developing and testing these models is tedious, painstaking work, but it is illustrative of the care that goes into serious scientific research. Most of it is not glamorous, and most scientists will go their whole careers without revolutionizing their disciplines or addressing the public.

These models also serve as a helpful counterpoint to the modeling-on-the-cheap approach taken by the anti-evolutionists. But we must attend to one more bit of table setting before coming to that.

5.4   THE ART OF COUNTING

Counting is simple when we have a small number of objects laid out sequentially before us. We just point to the objects in turn and say, "One, two, three, ..." We teach small children how to do that. But counting can become surprisingly difficult in more abstract settings.

The branch of mathematics devoted to counting is called "combinatorics," and we saw in Section 5.2 that it is closely related to probability. Going forward, we will need a few basic combinatorial principles.

We have already helped ourselves to one such principle: If we have n ways to do one thing, and k ways to do another, then there are nk ways of doing both together. For example, since each die has 6 possible outcomes by itself, the total number of outcomes for two dice is $6 \times 6 = 36$. Had we rolled a six-sided die and a twenty-sided die at the same time, there would have been $6 \times 20 = 120$ possible outcomes.

This formula has an interesting consequence. Suppose we ask for the number of ways of arranging the six letters D, A, R, W, I, N. One such arrangement is "DARWIN," but there are many others: NIWRAD, DIRWAN, and AWINRD, for example. If we listed all such arrangements, how long a list would we have?

We can use our basic principle like this: Arranging the letters is a six-step process. There are six ways of carrying out the first step, since any of the six letters could appear first. Once we have chosen the first letter, any of the remaining five can be the second letter. So there are five ways of carrying out the second step. Continuing in this manner, the total number of ways of ordering the letters is:

$$6 \times 5 \times 4 \times 3 \times 2 \times 1 = 720.$$

There was nothing special about using six letters. If instead we had n distinct letters we would have multiplied all the numbers from n down to 1. This multiplication is written as $n!$ and is called "n factorial." The exclamation point indicates our surprise at how fast these numbers grow. For example, $10!$ is already more than three million.

Let us try something more complex. We have seen that the number of ways of ordering n distinct letters is $n!$. What happens if the letters are not distinct? For example, how many ways can we order the letters E, V, O, L, U, T, I, O, N? Obviously, EVOLUTION is one possible ordering, but, again, there are many others.

To answer our question, suppose that we could distinguish the two occurrences of the letter o. We could imagine using a capital O in one place and a lowercase o in the other. Then we would have 9 different letters and the answer would just be 9!. Let us call this the "modified problem," since we have modified the original problem to obtain this answer. We now ask the following question: How should we adjust the answer to the modified problem to account for the fact that in our original problem the two occurrences of o cannot be distinguished?

The difference between the two problems is this: In the modified problem, the strings EVOLUTION and EVOLUTION are counted as different, but in the original problem they are the same. That means every string in the original problem gets counted twice when we shift to the modified problem. Since the answer to the modified problem was 9!, the answer to the original problem must have been 9! /2.

Now we move on to the final exam. How many ways can we arrange the letters in the word MISSISSIPPI?

This is trickier than our other examples, but the same logic applies. If we modify the problem so that all 11 letters are distinguishable, then the answer is 11!. Alas, the modified problem counts MISSISSIPPI as different from MISSISSIPPI, whereas the original problem counts them as the same. Thus, just as in the previous example, we need to divide by 2.

The modified problem also treats the four occurrences of s as different from one another. There are 4! = 24 ways of arranging these four symbols. That means that every sequence in the original problem gets counted 24 times in the modified problem, implying that after we divide by 2, we still have to divide by 24. We have to do this again to account for the 24 ways of ordering the four occurrences of I.

Bringing everything together, and keeping in mind that 2! = 2 and 4! = 24, the answer to the original problem is

$$\frac{11!}{(4!)(4!)(2!)}.$$

If you are curious, that works out to be 34,650, but the precise number is not really what is important. It is the reasoning underlying the answer that we care about.

5.5   THE BASIC ARGUMENT FROM IMPROBABILITY

We are now ready to see how anti-evolutionists use probability theory in their arguments.

We have noted that they start with a track one argument that evolution is asking us to believe that *something* very improbable has occurred. Their various track two arguments are then attempts to state precisely what that something is. For this purpose, they often take aim at the specificity of proteins, and we will consider some of their efforts in this regard in the present section.

We start with an especially clear example, written by David Foster, an ID proponent. His argument involves hemoglobin, a protein that transports oxygen through the blood of humans and other vertebrates.

Hemoglobin is a complex molecule possessing an elaborate, three-dimensional structure. However, Foster's argument treats it as a simple string of 574 amino acids. This is a crude oversimplification, but we can accept it for the sake of argument. Our goal is to expose the fallacious mathematical reasoning underlying the argument, as opposed to obsessing about every biological detail.

There are twenty kinds of amino acids in total. Foster begins his argument by listing the frequency with which each amino acid appears in hemoglobin. For example, the amino acid glycine appears 36 times, alanine appears 68 times, and so on. He now writes:

> The specificity of hemoglobin is described by the improbability of the specific amino acid sequence occurring by random chance.
> Such specificity is capable of exact calculation in the permutation formula:

$$P = \frac{N!}{n_1! \times n_2! \times n_3! \ldots \text{etc.}}$$

where N is the total number of amino acids in hemoglobin (574); n_1, etc. are the number of separate kinds of amino acids; and ! means that the given separate numbers are subjected to "factorial" expansion. Thus: $5! = 5 \times 4 \times 3 \times 2$.   *(Foster 1999, 80)*

That centralized formula is precisely the one we derived at the end of Section 5.4. Foster is discussing the "specificity of hemoglobin" in precisely the same way we described ordering the letters in MISSIS-SIPPI. He now continues:

> In the case of hemoglobin, ... the specific numerical value of the solution is $P = 10^{654}$. This is an immense number, 10 multiplied by itself 654 times. ... Thus we can state that the improbability of hemoglobin occurring by random selection can be represented by the infinitely small number 10^{-654}, which means 10 divided by itself 654 times: as near to zero as one could consider. ... This raises the question as to whether such very low probabilities are of a miraculous nature when they occur in factual situations such as the protein hemoglobin – whether such extremely improbable events are relevant to the question "Does God exist?"
>
> *(Foster 1999, 80)*

For the sake of accuracy, we should note that 10^{-654} does not actually mean to divide 10 by itself 654 times. Rather, it means to divide 1 by 10 a total of 654 times. This is equivalent to dividing 10 by itself 653 times.

Here is another version of the same argument, put forth by Ariel Roth, a young-Earth creationist:

> Suppose we need a specific kind of protein. What are the chances that amino acids would show up in the specific order required? The number of possible combinations is unimaginably great, because there is the possibility of any one of 20 amino acids occupying each position. For a protein needing 100 specific amino

acids, the number has been estimated at many times greater than all the atoms in the universe. Hence the chance of getting a necessary kind of protein is extremely small. ... For 100 specific amino acids the chance of getting the right kind of protein is only 1 out of the number 49 followed by 190 zeros (4.9×10^{-191}). Other similar calculations yield numbers that are also beyond the realm of plausibility.   *(Roth 1998, 69–70)*

Arguments of this general sort are ubiquitous in anti-evolutionist literature. Foster's article was published in a mainstream American periodical called *The Saturday Evening Post*, and I am sure its generally conservative readership was delighted by the superficial precision of his argument. I used Roth's example because he presented an easily quotable version of the argument, but most young-Earth creationist books present some version of it.

Regardless of the precise variation, the underlying logic of the argument is always the same:

1. Identify a complex biological structure, such as a specific gene or protein.
2. Model its evolution as a process of randomly selecting one item from a very large space of equiprobable possibilities.
3. Use elementary combinatorics to determine the size of the space, which we shall call S.
4. Conclude that the probability of the structure having evolved by chance is $1/S$, and assert that this is too small for evolution to be plausible.

I shall refer to this as the Basic Argument from Improbability (BAI). What can be said in reply?

Let us recast the BAI in the language of Section 5.2. We will assume the argument is being applied to a protein. In this case, our probability space consists of the set of proteins of a certain length. The BAI equips this space with the uniform distribution, meaning, recall, that we assume that any individual protein is as likely to occur as any other. Evolving a specific protein is then modeled as selecting one point at random from this space. Since the space is vast, the number of attempts needed to achieve success is prohibitive, even given geological time scales with which to work. At this point you assert that evolution has been disproved and call it a day.

However, you will have noticed that a critical part of the evolutionary process has been left out of this model. Biologists do not claim that the proteins found in modern organisms arose in one step by random selection from the space of all theoretical possibilities. The actual claim is that they arose in a gradual, step-by-step manner, as random variations were passed through the sieve of natural selection. Moreover, as we have noted, the problem is never to explain how some modern protein arose from scratch, but is rather to explain how the modern protein evolved from some more ancient protein, one that was possibly simpler and more rudimentary than the modern form.

Recall also our discussion from Section 4.2. Evolution finds its ultimate starting point at the origin of life, which we can think of as situating life at some specific point in protein space. As the process plays out, we do not search protein space as a whole, but instead carry out a sequence of local searches in the neighborhood of the point at which we happen to find ourselves. We noted that the vastness of the space as a whole is irrelevant because we only search a tiny fraction of it. This is illustrated in Figure 5.2.

We can now point to the first serious error with the BAI: it equipped its probability space with the uniform distribution, which is terribly unrealistic. Parents are very unlikely to have offspring with a protein far away from theirs. It is much more likely that the offspring has a protein that is local to the parents. Therefore, proteins near to our starting point are far more likely to be sampled than are proteins that are far away. The correct distribution would assign a probability close to 0 for most of the space – since most proteins will be very far away from our starting point – and then a very high probability to proteins that are nearby.

This leads us to the second serious problem with the BAI: it ignores the role of natural selection in affecting the probability of finding certain proteins. Suppose that some nearby proteins represent improvements over our current state, while other nearby proteins are

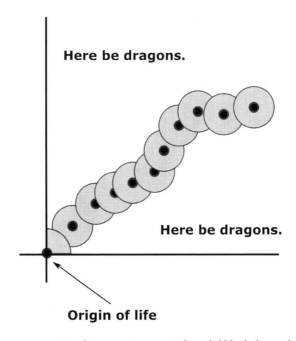

FIGURE 5.2 Searching protein space. The solid black dot at the intersection of the axes represents the origin of life. The other solid black dots represent protein stepping stones connecting an ancient protein to more modern forms. The gray circles show that evolution only examines the local area near an already existing protein and completely disregards most of the space.

harmful. Natural selection will then ensure that the improvements are more likely to be represented in future generations than are the harmful proteins. Mathematically speaking, we could say that the effect of selection is to dramatically shift the probability distribution toward favorable proteins and away from harmful ones. If there is a path of steady improvement connecting the rudimentary protein to its modern descendant, then natural selection will preserve our progress along that path. It will prevent us from moving backward while we wait for the next improvement.

We can dramatize these issues with an analogy. Suppose you and a friend are in the downtown area of a major American city, and you both decide you want a slice of pizza. You pick a direction and start walking. Within just two blocks you find a pizza parlor. Your friend now says, "Incredible! The surface of the Earth is enormous, and almost none of it is covered with pizza parlors. Yet somehow we were able to find one of the few places on Earth that has a pizza parlor. How can you explain something so remarkable?"

In this context, the error is obvious. The surface area of the earth is irrelevant because we only needed to search the tiny portion of it near our current location. And while pizza parlors are rare on the surface of the earth generally, they are extremely common in the downtown areas of major American cities. The BAI is guilty of precisely the same oversights, except applied to protein space rather than to the surface of the earth.

A more recent version of this fallacy occurs in the book *Darwin's Doubt*, written by ID proponent Stephen Meyer. Referring to the Wistar conference we discussed in Chapter 4, he writes:

> And that was the problem, as the Wistar skeptics saw it: random mutation must do the work of composing new genetic information, yet the sheer number of possible nucleotide base or amino-acid combinations (i.e. the size of the combinatorial "space") associated with a single gene or protein of even modest length rendered the probability of random assembly prohibitively small. For every sequence of amino acids that generates a functional protein, there are a myriad of other combinations that don't. As the length of the required protein grows, the number of possible amino-acid combinations mushrooms exponentially. As this happens, the probability of ever stumbling by random mutation onto a functional sequence rapidly diminishes.
>
> *(Meyer 2013, 173)*

The logic underlying this argument fits our bullet-point list perfectly, and Meyer's argument fails for the reasons we have already enumerated. The probability space implied by his argument is entirely unrealistic biologically since it includes no role for natural selection.

However, unlike Foster and Roth, Meyer is too sophisticated to ignore natural selection altogether. Just prior to our previous quote, he presents his reason for discounting it:

> Clearly, natural selection plays a crucial role in this process. Favorable mutations are passed on; unfavorable mutations are weeded out. Nevertheless, the process can only select variations in the genetic text that mutations have first produced. For this reason, evolutionary biologists typically recognize that mutation, not natural selection, provides the source of variation and innovation in the evolutionary process. As evolutionary biologists Jack King and Thomas Jukes put it in 1969, "Natural selection is the editor, rather than the composer, of the genetic message."
>
> *(Meyer 2013, 173)*

This is a remarkable paragraph. One would have thought it obvious that mutation and selection both play a crucial role in the process of evolutionary innovation. According to evolutionary theory, complex, functional structures arise through the interplay of mutation and selection. It does not make sense to say that one or the other is the ultimate source of the adaptation. It is fine to say that mutation provides the ultimate source of *variation*, but it is flatly wrong to say it is the source of *innovation*. Continuing with the analogy of writers and editors, mutation is like a sloppy, self-indulgent writer who buries a few pages of insightful writing within hundreds of pages of unpublishable nonsense. The editor then plays a crucial role in the process by separating the few worthy pages from the mountains of chaff.

Consider an analogy. Imagine a child learning to play chess. He plays many practice games with his coach, who dutifully critiques his moves and shows him better ways of playing. Years later the child becomes a master-strength player. Meyer's argument is tantamount to saying that the coach played no role in the process of the student becoming a master. After all, the coach had nothing to critique until the child actually made some moves at the board. In the analogy, mutation is like the child making moves and natural selection is like the coach passing judgment on them. And just as the child's moves and the coach's commentary were both essential to turning the child into a master, so too are mutation and selection both essential to the process of biological innovation.

Summarizing, we have here another instance of evolution's critics failing to take into consideration the probabilistic and geometric structure of protein space. Specifically, the probability distribution appropriate to protein space is highly nonuniform, meaning that some proteins are far more likely to be found in organisms than others. And our movement through protein space is highly constrained by its geometric structure since each generation almost exclusively investigates points that are near to the generation before. The BAI ignores both of these points and can be dismissed for these reasons.

There is also a third serious problem with the BAI: It treats low probability, all by itself, as a refutation of evolution and confirmation of design. The problem with this approach is that extremely improbable events happen all the time. We have previously mentioned the old saying that million to one odds happen eight times a day in New York City. The biological analog is this: The course of evolution is affected by so many chance events that any specific modern outcome of the process could be extremely low, but this does not make us suspect design, because *something* had to happen. Therefore, some additional argument is needed to go from low probability to a conclusion of design. We need to argue not simply that, starting from the origin of life, a given protein was unlikely to have evolved, but also that this is the sort of improbability that needs a special explanation. Proponents of the BAI do not supply that additional argument.

The BAI has the great virtue of computational tractability. By reducing probability to combinatorics, its proponents are able to carry out actual calculations to produce the desired small numbers. However, we have seen that the mathematical model on which the BAI relies is far too unrealistic to produce meaningful results. Evolution does not proceed by randomly selecting complex structures from spaces of equiprobable possibilities, but instead builds them gradually by accumulating many small changes, a process we can refer to as "cumulative selection."

Including a role for cumulative selection in the model would make it all but impossible to carry out meaningful calculations. The number of variables influencing the probabilistic structure of protein space is enormous. It is effectively impossible to identify them all, much less to assign appropriate numbers to them.

Thus, anyone seeking to use probability theory to refute evolution must deal seriously with these two questions:

- How do we preserve the computational tractability of the BAI while working within a biologically reasonable model?
- How do we justify the conclusion that evolution has been refuted simply because a particular mathematical model tells us that an event of low probability has occurred?

The remainder of this chapter will discuss the main attempts from ID proponents to circumvent these issues. As we shall see, they are not successful in doing so.

5.6   COMPLEX SPECIFIED INFORMATION

If you spend much time reading ID literature, you will encounter the concept of "complex specified information (CSI)." In the form used by ID proponents, the concept is the brainchild of mathematician William Dembski. In his telling, CSI is an attribute that an event may or may not exhibit. We are using the term "event" very broadly to refer to any particular occurrence or object. He claims that when CSI is present, you can be certain that intelligent design is in some way involved in the causal history of the event. He writes:

> When trying to explain something, we employ three broad modes of explanation: *necessity, chance* and *design.* As a criterion for detecting design, specified complexity enables us to decide which of these modes of explanation apply. It does that by answering three questions about the thing we are trying to explain: Is it

contingent? Is it complex? Is it specified? By arranging these questions sequentially as decision nodes in a flowchart, we can represent specified complexity as a criterion for detecting design.

(Dembski 2004, 87)

In this scheme, "complex" indicates that the event occurs with low probability, while "specified" indicates that the event conforms to some independently describable pattern. Dembski's basic idea is readily explained through a few examples.

Suppose we flip a coin 20 times and obtain this sequence of heads and tails:

$$H\,H\,T\,T\,T\,H\,T\,H\,T\,T\,T\,H\,H\,H\,T\,H\,T\,H\,T\,T$$

How likely were we to get this exact sequence? Using what we learned in Section 5.2, we would say that each toss is independent of the other tosses, meaning we can multiply the individual probabilities. Since each toss comes up H or T with probability $1/2$, the probability of this sequence will be $1/2$ multiplied by itself 20 times. That works out to less than one in a million, which for illustrative purposes we can take to represent a low probability. However, this low probability by itself does not make us suspect trickery, since events of low probability occur all the time. This exact sequence was extremely unlikely, but *something* had to happen.

Now suppose that we got this sequence instead:

$$H\,H\,H\,H\,H\,H\,H\,H\,H\,H\,H\,H\,H\,H\,H\,H\,H\,H\,H\,H$$

This exact sequence has precisely the same probability as before: less than one in a million. But now we have an obvious pattern as opposed to some random jumble of heads and tails. In Dembski's terminology, we now have both complexity and specificity, and this makes us suspect some sort of trickery. Perhaps the coin was tossed by a skillful sleight of hand artist who was subtly manipulating how the coin would land, or perhaps the coin was weighted in a way that made it almost inevitable that it would land heads. We would suspect an explanation of that nature, rather than think the event occurred from a fair coin flipped in a fair way.

We could make the same point with a deck of cards. Shuffle the cards and deal them out on a table. Reasoning as with our coin tosses, the probability of getting that exact sequence of cards was very small, 1/52! to be exact, but we are not yet suspicious, because *some* sequence of cards had to appear.

Looking now at the entire 52-card sequence, we might find, say, two consecutive aces or three consecutive hearts. Those short runs have specificity, but we do not have complexity, since we expect short patterns like that to happen by chance with high probability. But if the 52 cards have sorted themselves by suit, with each suit running in sequence from ace to king, then we have both complexity and specificity, and again we suspect some sort of trickery.

We could also imagine throwing a large number of lettered tiles onto a table. We would not be surprised if simple two-letter or three-letter words appeared just by chance. They are specified, in the sense that they are recognizable as English words, but they are not complex, since a few short strings of that kind will occur by chance with high probability. A lengthy, nonsensical string of letters might be complex, but since it is not specified we do not yet suspect design. But if the tiles form a long English sentence, then we have both complexity and specification, and we once again suspect some sort of trickery. If a friend told us he did this experiment and that the tiles just happened to spell out "It was a dark and stormy night," then we would suspect he was not telling us the whole story.

This sort of reasoning arises in many situations. Any pattern of crags and grooves on a mountain is extremely unlikely, but the faces on Mt. Rushmore are also specified, and therefore we quickly identify them as having arisen from design. An archaeologist can quickly distinguish stone tools from random rock formations, since the former exhibit arrangements of parts that are unlikely to occur by chance, and since the arrangement is specified by virtue of being useful for a clear purpose. A large number of metal gears and springs can be arranged in many ways, but if we are confronted with an arrangement that makes a functioning pocket watch, then we know we have an instance of intelligent design. We could multiply such examples endlessly.

Dembski first presented these ideas to the public in his 1998 book *The Design Inference* (Dembski 1998). To this point, the argument has nothing to do specifically with biology. Dembski's intention was to provide a rigorous and general method for detecting whether design was involved in the causal history of any event, whether drawn from nature or from human history. After presenting his framework, he goes on to argue that he can apply it to biology. He claims that we can take for our event the appearance of some complex biological system and apply his method to it. When he does so, he continues, the result amounts to a mathematical proof that the system is in some way the result of intelligent design.

For example, we might take as our event a complex structure like the flagellum used by some bacteria to propel themselves. It is a complex biomolecular machine that involves numerous proteins working in concert. Following Dembski's lead, we might say, "The probability of that specific arrangement of parts arising by chance is extremely small, and the flagellum is specified by virtue of the fact that it is a functional machine that operates in much the same way as an outboard motor on a boat. The flagellum is therefore both complex and specified, and this implies that it cannot be explained without some recourse to intelligent design." As it happens, the flagellum is the main example Dembski uses to make his case for ID, as we shall discuss in Section 5.7.

Though it will not be relevant to the ensuing discussion, we should note for completeness that Dembski puts forth 10^{-150} as the probabilistic threshold for complexity. In other words, he argues, based on various criteria drawn from physics, that an event with a probability smaller than one in 10^{150} should be considered complex for the purposes of his framework.

At this point we should recall once more the parallel tracks of mathematical reasoning. Taken at a track one level, there is a superficial plausibility to Dembski's argument. You were probably nodding along with the foregoing examples, thinking, "Why, yes, come to think of it, improbable things that fit a pattern really do suggest intelligent design." However, if we are to accept this as a serious argument, then we also need the second track, where we pass from general intuitions to rigorous mathematics. Dembski claims his methods allow him to prove mathematically that evolution has been refuted, but there are several issues to resolve before we can analyze that claim.

It is one thing to apply Dembski's method to simplistic examples involving coins and playing cards, where it is easy to carry out probability calculations, but everything becomes murky when we consider nontrivial examples, especially those drawn from biology. If the causal history of an event is entirely unknown to us, then how can we carry out a meaningful probability calculation for its occurrence? As we have seen, any such calculation requires that we define a probability space. Doing so entails knowing the full set of alternatives in which the event is embedded and the probability distribution appropriate to those alternatives. However, possessing that knowledge would seem to require some familiarity with the causal history of the event.

"Specification" is likewise problematic, since there is a danger of doing the equivalent of looking at a fluffy cumulus cloud and saying, "Gosh, that looks sort of like a dragon." We have enough experience with coins and playing cards to distinguish design-suggesting patterns from more mundane arrangements. We know what mountains look like when we do not carve faces into them, and that makes it easy to recognize Mt. Rushmore as something designed. This background knowledge is precisely what we lack in the biological context. No one has intuitions about what will arise after billions of years of evolution starting from a relatively simple sort of life. We have no base of experience allowing us to say, "This structure is not the sort of thing evolution can produce, so we will have to explain it by recourse to intelligent design." Thus, we need to be rigorous about how we distinguish the design-suggesting patterns from the ones we impose on nature through excessive imagination, and it is very unclear that this can be done in general.

There are also difficulties with how complexity and specification relate. Dembski's method tells us first to carry out a calculation to establish complexity and only then to consider the question of specification. It is not clear that this is workable, since the manner in which we specify the event will influence the probability space we use to carry out the calculation.

In light of these and many other issues, we have reason to be skeptical of Dembski's framework as a general method for detecting design. At a minimum, when Dembski attempts to apply his method to biology, we will need to pay close attention to how he tries to circumvent the difficult questions we have raised.

We shall see that when the object in question is a biological adaptation, Dembski has no sound way of establishing either complexity or specification in the precise technical sense that his theory requires.

5.7 IS THE FLAGELLUM COMPLEX AND SPECIFIED?

In all of Dembski's voluminous writings defending both the theoretical rigor and practical utility of his ideas, he has only once tried to apply his method to an actual biological system. This occurred in his 2002 book *No Free Lunch: Why Specified Complexity Cannot Be Purchased Without Intelligence.* For his target he chose the flagellum of the bacterium *E. coli.* As we have mentioned, the flagellum is a whip-like appendage that gyrates back and forth, thereby propelling the bacterium through its environment. It is composed of numerous coordinated proteins that operate in a manner reminiscent of an outboard motor on a small boat. For his argument to work, Dembski needs to show the flagellum is both complex and specified.

Within Dembski's framework, establishing complexity requires a probability calculation. He bases his calculation on the concept of irreducible complexity we discussed back in Section 2.5. Recall that an irreducibly complex system is one that requires several mutually dependent parts to function. That is, the removal of any one part causes the machine to fail catastrophically. We noted that this concept was introduced by Michael Behe, who argued that an irreducibly complex system could not evolve in a gradual, stepwise manner.

Dembski accepts this argument, writing:

> Richard Dawkins has memorably described this gradualistic approach to achieving biological complexity as "climbing Mount Improbable." ... For irreducibly complex systems that have numerous diverse parts and that exhibit the minimal level of complexity needed to retain a minimal level of function, such a gradual ascent up Mount Improbable is no longer possible.
>
> *(Dembski 2002, 290)*

He then argues:

> An irreducibly complex system is a discrete combinatorial object. Probabilities therefore naturally arise and attach to such objects. Such objects are invariably composed of building blocks. Moreover, these building blocks need to be [*sic*] converge on some location. Finally, once at this location the building blocks need to be configured to form the object. It follows that the probability of obtaining an irreducibly complex system is the probability of originating the building blocks required for the system, multiplied times the probability of locating them in one place once the building blocks are given, multiplied times the probability of configuring them once the building blocks are given and located in one place.
>
> *(Dembski 2002, 290–291)*

Thus, there are two parts to the calculation:

- Use irreducible complexity to justify treating the system as a "discrete combinatorial object."

- Model the system's existence as the result of a threefold process of origination, localization, and configuration.

At the end of Section 5.5, I presented two questions that any anti-evolution probability argument must answer: Why is it legitimate to reduce probability to combinatorics, and why should evolutionists worry about the small number at the end of the calculation? We now see how Dembski proposes to answer these questions:

- Irreducibly complex systems cannot evolve in a gradual, stepwise manner, and this justifies treating them as "discrete, combinatorial objects."
- Functional systems are not only improbable, but also specified, and it is the combination of the two that explains why we should infer design.

Now, in Section 2.5, I used Michael Behe's original definition of irreducible complexity, as presented in his 1996 book. Most recent ID writing in the present employs precisely this definition. However, Dembski developed his own definition, in an attempt to circumvent the sorts of criticisms I discussed in Section 2.5. Here is that definition:

> A system performing a given basic function is *irreducibly*
> *complex* if it includes a set of well-matched, mutually interacting,
> nonarbitrarily individuated parts such that each part in the set is
> indispensable to maintaining the system's basic, and therefore
> original, function. The set of these indispensable parts is known
> as the *irreducible core* of the system.   *(Dembski 2002, 285)*

This definition is only reasonable if we accept that the "basic" function of a set of parts in the present must be the same as its original function in some precursor system. There is no reason at all to accept that, however, since we have already seen that cooption of function is commonplace in evolution. This is especially relevant when considering the flagellum, since there is substantial evidence that such cooption occurred in the course of its evolution.

If we eliminate the clause "and therefore original," then this definition is only trivially different from Behe's, and the criticisms we made against his argument apply with equal force here. Mutual interdependence of parts in the present has no relevance at all to the possibility of functional precursors in the past, and that is true regardless of whether we use Behe's definition or Dembski's definition.

Let us return now to Dembski's flagellum calculation. He justifies his reduction of probability to combinatorics by asserting that an irreducibly complex system cannot emerge by gradual evolution. Since we have seen that this is false, the logic underlying the calculation collapses. Dembski's use of probability is therefore no improvement at all over the BAI. We can reject his argument without considering the details of his origination-localization-configuration model for the appearance of a complex adaptation. (As it happens, those details will be relevant in Section 5.8, so we will defer further discussion until then.) Thus, Dembski has no way to establish complexity in the precise sense that he needs.

Let us see if he does any better in establishing specification. He writes:

> Biological specification always refers to function. An organism is a functional system comprising many functional subsystems. In virtue of their function, these systems embody patterns that are objectively given and can be identified independently of the systems that embody them. *(Dembski 2002, 148)*

With regard to the flagellum specifically, he writes:

> [I]n the case of the bacterial flagellum, humans developed outboard rotary motors well before they figured out that the flagellum was such a machine. This is not to say that for the biological function of a system to constitute a specification humans must have independently invented a system that performs the same function. Nevertheless, independent invention makes the detachability of a pattern from an event or object all the more stark. At any rate, no biologist I know questions whether the functional systems that arise in biology are specified. *(Dembski 2002, 289)*

We can smile at that last remark, since Dembski's idiosyncratic version of "specificity" plays no role at all in contemporary biology.

This paragraph does little to allay our fear that saying of a flagellum that it resembles an outboard motor is comparable to saying of a cloud that it resembles a dragon. Biologists argue that evolution produces functional structures as a matter of course, and this lends some urgency to the question of whether the function of a system can serve as a specification in Dembski's sense. In other words, in the context of evolution, does bearing a resemblance to an outboard motor constitute a design-suggesting pattern, or is it just a normal outcome of the evolutionary process?

Dembski's notion of "detachability" is meant to address this issue. Roughly, he means that a proper specification is one that is describable without any reference to the object itself. A simple analogy that illustrates the idea is firing an arrow at the side of a barn. After firing the arrow, we can paint a small circle around its landing point to pretend that we hit a very small target. That would be a nondetachable pattern – we needed to see where the arrow landed before painting the target. But if we paint a small target on the barn before firing, then we have a detachable pattern.

A significant section of *No Free Lunch* is devoted to providing a mathematical justification for the notion of "detachability." This section is impressively technical, invoking difficult ideas from the theory of statistical hypothesis testing. First I will present the crucial part of the discussion precisely as Dembski presents it, and then we will discuss what it means:

> Given a reference class of possibilities Ω, a chance hypothesis H, a probability measure induced by H and defined on Ω (i. e. $P(\cdot \mid H)$), and an event/sample E from Ω; a rejection function f is *detachable* from E if and only if a subject possesses background knowledge K that is conditionally independent of E (i. e. $P(E \mid H\&K) = P(E|H)$) and such that K explicitly and univocally identifies the function f. Any rejection region R of the form $T^{\gamma} = \{\omega \in \Omega \mid f(\omega) \geq \gamma\}$ or

> $T_\delta = \{\omega \in \Omega \mid f(\omega) \leq \delta\}$ is then said to be *detachable* from E as
> well. Furthermore, R is then called a *specification* of E and E is
> said to be *specified*.   *(Dembski 2002, 62–63)*

This is written very much in the style of track two mathematical reasoning, so let us now try to translate this back into track one.

We can illustrate the idea with a classic experiment described by Ronald Fisher in his 1935 book *The Design of Experiments* (Fisher 1935). Apparently Fisher attended a party at which people were drinking tea with milk. One of the women at the party claimed that she could distinguish, by taste, a cup in which the tea was poured into the cup before the milk from one where the milk was poured before the tea. Fisher was skeptical and quickly designed an experiment to test this claim. Eight cups of tea were prepared, half with the milk poured first and half with the tea poured first. The woman was challenged to identify the four cups in which the milk was poured first.

The question is this: how many cups does she need to get right for us to conclude that she is not just getting lucky? Does she have to get all four? Or would we be impressed if she only got three or two of them correct?

Fisher reasoned that the key was working out the probability of getting a particular result by chance. If that probability is below a certain threshold, then we conclude that the result is not due to random guessing. Fisher took 5% as his threshold. In other words, he decided ahead of time that if the probability of getting a given number of correct answers by random guessing was below 5%, then he would conclude that the woman really had the ability she claimed to have.

In this context, the probability calculations are not too difficult. In fact, they can be carried out using only a small variation on the counting techniques we discussed in Section 5.4. Fisher found that the probability of getting all four right by random guessing was roughly 1.4%, below the threshold. On the other hand, the probability of getting at least three of them correct was roughly 24%. This probability is too high for us to discard random guessing as a viable hypothesis. The 5% threshold was arbitrary, but it seems to work well enough in practice.

We can quickly generalize Fisher's approach. Speaking informally, suppose we have collected some data, and we want to know if something other than pure chance is at work. We start by putting forth the "null hypothesis" that nothing beyond chance is happening. We then work out the probability of obtaining the data under the null hypothesis. If the probability is below a certain threshold, typically taken to be 5% or 1% in modern statistical practice (at least in the social sciences), then we dismiss the null hypothesis as not credible. In statistics jargon, probabilities below the threshold are said to fall into the "rejection region." Applying this method requires that we have a grasp on the appropriate probability distribution to use, but this is possible in many practical situations.

Return now to our very technical quotation of a few paragraphs ago. Dembski's intent is to provide a rigorous mathematical foundation for his notion of "specification." He attempts this by adding a minor gloss to Fisher's approach to hypothesis testing. The first part of the quote just says that we have a set of possible outcomes equipped with a probability distribution and that we have collected some empirical data. We want to know if the data fits a pattern that suggests design. In Dembski's framework, we do this by using a function to assign a number to each outcome and then by determining whether our data corresponds to an extreme value for that function. In the jargon of probability theory this function is known as a "random variable." To ensure we defined our random variable honestly, we require that it arise from background knowledge that is independent of any knowledge of the data itself. The rejection regions, which are represented by the notations T^γ and T_δ, represent values of the random variable that are unlikely to be due to chance alone.

Visually, this is shown in Figure 5.3. Most events correspond to points in the white region under the graph, and these are of sufficiently high probability that we regard chance alone to be a plausible explanation. The notations T^γ and T_δ refer to the tails of

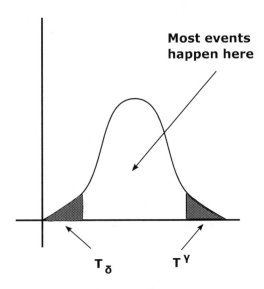

FIGURE 5.3 In a statistical experiment, some results can be attributed to chance, while others are deemed sufficiently improbable as to require some other explanation. Visually, we might say that most events are found in the white region under the graph. Those in the shaded zones are deemed to have fallen into the "rejection region," meaning that we reject chance alone as an explanation.

the distribution, shaded gray in the figure. Events in these "rejection regions" are deemed to be so improbable that chance alone is considered unlikely as an explanation.

We can readily present the "lady drinking tea" experiment in these terms. The set of possible outcomes contains all the ways the lady might have assigned the labels "tea first" or "milk first" to the eight cups. Our chance hypothesis is that she is just guessing, and this induces the uniform probability distribution on our set of possible outcomes. Our random variable assigns to each possible outcome the number of "tea first" cups she successfully identified. Since we can assign these probabilities without knowing the result of the experiment, this is a detachable function. Finally, the rejection region corresponds to getting all four "tea first" cups right, since that has a probability of less than 5%.

This might still seem very technical, but the underlying idea is pretty clear. If you specify a region of low probability before carrying out the experiment, and then the data ends up falling right into that region, you suspect that something other than chance is at play. Dembski refers to that low-probability region as a "specification." If the experiment has already been done, then it is important that the specification be describable without reference to the data itself. This is what Dembski means in saying that a specification has to be "detachable."

However, you will notice that Dembski provides nothing comparable to our discussion of the "lady tasting tea" experiment for his flagellum example. Have another look at that highly technical quotation. He defines various mathematical objects, represented by letters such as Ω, K, E, f, T^γ, and T_δ. When we apply Dembski's framework to an actual example, we need to be able to link up these letters to their real-world counterparts. Otherwise we have simply defined a lot of mathematical symbolism for no good reason.

In other words, early in the book, he devotes many dense, technical pages to developing a piece of mathematical machinery said to identify design-suggesting specifications. But when it comes time to apply this machinery later in the book, it is as though these earlier pages never happened. Dembski just declares it obvious that the flagellum is specified because it resembles an outboard motor. He does not explain how our background knowledge "explicitly and univocally" defines an appropriate rejection function, nor does he quantify anything that might help us understand his sets T^γ or T_δ.

As soon as you try to apply Dembski's machinery to the flagellum you realize how hopeless it is to assign values to any of his variables. The space of possibilities, represented by Ω, would presumably have to be all the possible bacterial genotypes that might have evolved in place of the one that did, but, as we have discussed, we have no idea at all what probability distribution to use for that space. We also have no relevant background knowledge that would help us define a useful rejection function, represented by f. Moreover, if we argue that the flagellum is specified by virtue of its function, then the rejection regions, represented by T^{γ} and T_{δ}, would have to consist of all the possible propulsion systems that might have evolved and not just the particular flagellum that did. How on earth do you propose to define that region with the necessary precision?

Therefore, the situation with regard to the specification of the flagellum is this: We started with a vague notion that the flagellum looks a bit like an outboard motor, but we worried that asserting such a thing might be tantamount to looking at a cloud and seeing a dragon. Dembski promised us a rigorous mathematical test for identifying design-suggesting patterns, and sure enough his book contains many dense, technical pages, filled with jargon and notation, purporting to do just that. But this machinery is never heard from again. When Dembski actually discusses the flagellum, he just says it is obviously specified because it looks like an outboard motor.

Dembski's framework for inferring design requires that the flagellum be both complex and specified. But we have no way of establishing complexity because we cannot carry out a relevant probability calculation, and we have no way of establishing specificity because we have no background knowledge to help us distinguish the design-suggesting patterns from the ones arising from excessive imagination. This is an instance of a mathematically technical track two argument lacking a corresponding track one explanation to make it all comprehensible.

Therefore, Dembski's argument involving "complex specified information" is no improvement at all over the BAI.

5.8   GREEK LETTERS AND SUBSCRIPTS DO NOT HELP

Dembski published his flagellum calculation in 2002. Since the calculation was plainly based on fallacious assumptions, most scientists ignored it.

That might have been the end of the story, but for a paper that appeared in the *Journal of Theoretical Biology* (JTB) in late 2020. JTB is a reputable academic journal, the sort of venue that publishes research articles for an intended audience of professional scientists. In this case, however, the journal's editors seem to have erred by allowing a very poor paper to be published.

The article's authors, Steinar Thorvaldsen and Ola Hössjer, revived Dembski's calculation as part of a broader argument for finding evidence of ID in biology. They write:

> [D]embski proposes an equation based on three independent events: A_p: originating the building blocks (protein chains) of the protein complex....A_ℓ: localizing the building blocks in the same place, and A_c: configuring the building blocks correctly to form the complex. Then the probability of a protein complex is the multiplicative product of the probabilities of the origination of its constituent parts, the localization of those parts in one place, and the configuration of those parts into the resulting system (contact topology). This leads to the following estimate for the probability of a protein complex (PC) composed of N independent building blocks:
>
> $$\hat{P}(A_{PC}) = \prod_{n=1}^{N} \left[P\left(A_p^{(n)} \mid \hat{\theta}_p^{(n)}\right) \cdot P\left(A_\ell^{(n)} \mid \hat{\theta}_\ell^{(n)}\right) \cdot P\left(A_c^{(n)} \mid \hat{\theta}_c^{(n)}\right) \right],$$
>
> where $\theta_p^{(n)}$, $\theta_\ell^{(n)}$, and $\theta_c^{(n)}$ are the parameters involved in forming the protein chain, the localization and the configuration of the nth building block. *(Thorvaldsen and Hössjer 2020, 7)*

Here we have another instance of an attempted track two argument. Mathematical notation can be intimidating, and a technical equation like the one in the quotation can create the illusion that something profound has been said. However, it is important to keep in mind that mathematical symbols are just abbreviations, and equations are just sentences. As we have noted, when confronted with a pile of notation, it is important to pause for a moment to ask for the track one interpretation.

In this case, the equation is nothing more than a translation into symbols of Dembski's argument from Section 5.7. In other words, you can express the argument in plain English, as in Section 5.7, or you can first translate the English words into symbols and then express the same argument in the form of an equation. Looking at their equation, the capital Greek letter pi is a standard mathematical notation to indicate we are multiplying several things together. The three terms on the right hand side, each one containing a P followed by an expression in parentheses, represent the probabilities of origination, localization, and configuration put forth in the paragraph above the equation. The expression on the left hand side represents the probability of the particular protein complex in which we are interested – the flagellum in the case of Dembski's original argument. Loosely translated, the equation just says that the probability of the protein complex is found by multiplying together various other probabilities related to its component parts, precisely as the paragraph asserted in plain English.

Of course, mathematicians do this sort of thing all the time. We frequently translate natural language sentences into symbolic equations. Notation and equations make possible a level of precision that is difficult to achieve in a natural language. But then we typically go on to manipulate the equations, so as to learn something about the real-world objects to which the variables relate. Had Thorvaldsen and Hössjer gone on to use this equation for some practical purpose, it would not be necessary to belabor things as I have.

However, they do not go on to use this equation for anything. Continuing from where the previous quotation left off, they write:

> Modeling the formation of structures like protein complexes via this three-part process of production, convergence, and assembly, is of course problematic because the parameters in the model are very difficult to estimate. ... [T]he usefulness of the equation is not in the solving, but rather in the contemplation of all the various concepts which science must incorporate when considering the question of how to explain this kind of complex structures [sic].
> (Thorvaldsen and Hössjer 2020, 7)

In light of this very frank admission, we might wonder what purpose is served by the Greek letters and subscripts. Contemplating the various concepts is more easily done when those concepts are expressed in accessible language.

That said, are they at least right that the equation clarifies the issues involved in crafting a complex adaptation? Sadly, they are not. In Section 5.7, we noted that Dembski's flagellum calculation could be dismissed out of hand because it was based on the fallacious assumption that irreducible complexity posed a challenge for evolution. Let us put that aside now and consider whether there is any merit to his three-step model of flagellum assembly.

If we were modeling the construction of a building, then it might make sense to think in terms of origination, localization, and configuration. We start by ordering the construction materials from various suppliers, arrange to have those materials transported to the building site, and then configure the materials into a building. For a complex biological structure, however, this approach makes no sense at all.

Thorvaldsen and Hössjer, like Dembski before them, treat the assembly of a flagellum as primarily a combinatorial problem about arrangements of proteins. In reality they should be thinking in terms of genetic instructions. These instructions do not operate in three distinct phases of origination, localization, and configuration. Rather, the same instructions that dictate which proteins get produced also direct those proteins to specific locations and mandate the relationships among them. It is a package deal, so to speak.

This leads us to a further problem with the equation. From our discussion in Section 5.2, we know that whenever we see probabilities being multiplied we must ask whether they are independent of each other. Thorvaldsen and Hössjer assert that they are, but this is plainly not correct. Having appreciated that the issue is not about combinatorial arrangements of proteins, but is instead about genetic instructions, we understand that origination, localization, and configuration are not independent processes. Thus, even if we were to grant, for argument's sake, that we can usefully model the assembly of a flagellum as a three-step process, it still would not make sense to multiply the individual probabilities.

Mathematicians typically strive for the utmost clarity and precision in their writing. If your experience of math comes primarily from tedious high school classes, then you may find that hard to believe, but it is true. Used properly, notation and equations permit a level of precision and a clarity of thought that is not otherwise possible. Some practice and training is required to comprehend it all, but the fact remains that good mathematical writing is some of the clearest writing you will ever encounter.

In contrast, there is a prominent subset of ID literature that features copious mathematical jargon and notation whose purpose has little to do with bringing clarity to difficult subjects. In fact, we have now seen two examples where the intent seems to be precisely the opposite. Instead of bringing clarity, the intent appears to be to create an aura of complexity.

A bad argument does not become good when it is expressed with mathematical notation. Thorvaldsen and Hössjer's attempt to revive Dembski's calculation does nothing to address its many deficiencies.

We will have to look elsewhere for a probabilistic argument against evolution.

5.9   RECENT WORK ON FLAGELLUM EVOLUTION

Let us summarize the argument of the last few sections.

We have seen that ID proponents claim to have developed rigorous, mathematical tools with which they can detect intelligent design in the causal history of any event or object. These tools are based on the assertion that improbable things that fit a recognizable pattern can only arise from design. Furthermore, the argument continues, when these tools are applied to the bacterial flagellum the result is mathematical proof that it is in some way the result of intelligent design. In this way, ID proponents claim to have ruled out the possibility that the flagellum is the product of standard evolutionary mechanisms.

However, when we looked at the details we found that the argument failed at every point. Their probability calculation was both mathematically and biologically absurd, and we have no relevant background knowledge to help us determine whether or not the flagellum instantiates a design-suggesting pattern. We therefore rejected, and frankly scoffed at, the pretensions of ID proponents to mathematical seriousness. Anti-evolutionists have given us no good reason for thinking the flagellum could not have arisen through gradual evolution.

Can we do better from the other side? Arguing that there is no reason to reject evolution as the explanation for the flagellum is different from arguing that we have good reason for accepting evolution as the explanation. Do we have good evidence that the flagellum evolved, or should we just throw up our hands and lapse into agnosticism?

Over the last two decades, there has been extensive research into the flagellum, as well as into other, similar sorts of propulsion systems found in various microbes. Scientists who study these systems know a tremendous amount about the proteins comprising them and the manner in which such systems are assembled. This research has led to an unambiguous conclusion: the flagellum is the product of gradual evolution. It is not just that the flagellum might, in principle, have arisen through evolution, but rather that the evidence is very strong that it actually did so arise.

Careful study of the flagellum shows that it is not the handiwork of a master engineer, but is more like a cobbled-together mess of kludges. It shows the "senseless signs of history" we discussed in Section 2.4. In an article surveying this evidence, a team of researchers summarize the situation like this (note that "exaptation" is a technical term in evolutionary biology that in this case can be thought of as roughly synonymous with "cooption."):

> Many functions of the three propulsive nanomachines are precarious, over-engineered contraptions, such as the flagellar switch to filament assembly when the hook reaches a pre-determined length, requiring secretion of proteins that inhibit

transcription of filament components. Other examples of absurd complexity include crude attachment of part of an ancestral ATPase for secretion gate maturation, and the assembly of flagellar filaments at their distal end. All cases are absurd, and yet it is challenging to (intelligently) imagine another solution given the tools (proteins) to hand. Indeed, absurd (or irrational) design appears a hallmark of the evolutionary processes of co-option and exaptation that drove evolution of the three propulsive nanomachines, where successive steps into the adjacent possible function space cannot anticipate the subsequent adaptations and exaptations that would then become possible.

(Beeby et al. 2020, 290)

As is typical for articles of this sort, the jargon is formidable. However, the authors' point is precisely the one I emphasized in Section 2.4. Recall our discussion of the water pipes in my basement, or the arrangement of roads on the way to my father's office. In these systems everything worked well enough, but their organization only made sense when the historical process leading to them was taken into consideration. They embody patterns that no engineer would even consider, but which fall into place when you realize that each step made sense given what had come before.

The authors are saying the flagellum developed in the same way. If you examine its structure as a whole, then there are many aspects of it that do not make sense from an engineering standpoint. But these aspects make perfect sense when you consider the most likely sequence of steps through which the flagellum evolved. Each stage arose by modifying more rudimentary precursor systems or by coopting and repurposing already existing proteins, and this process left its marks on the finished, modern structure.

The story of the flagellum is a familiar one to biologists studying evolution. Anti-evolutionists make bold, sweeping claims that some complex system could not have arisen through evolution. They tell the world they have conclusive mathematical proof of this.

Scientists then expose the fatal conceptual flaws in the claimed proofs and present the extensive circumstantial evidence that the system actually arose through evolution. In response, the anti-evolutionists retreat to their fallback position, where they fold their arms, pout, and mutter that the evidence is insufficient to convince them.

We previously saw this same story for eyes. We could replay it again for numerous other complex systems the anti-evolutionists have occasionally championed. Invariably, the people who actually study these systems see clear evidence for evolution, not design.

5.10 THE EDGE OF EVOLUTION?

Anti-evolutionists have one other probability argument to offer, and we conclude this chapter by considering it.

Michael Behe has attempted a different approach to probability calculations, as presented in his 2007 book *The Edge of Evolution*. Unlike Dembski, or the proponents of the BAI, Behe does not apply his calculations to specific biological structures like gene sequences or bacterial flagella. Instead he identifies specific starting and ending points in protein space and seeks the probability of moving from one to the other.

Behe claims that the evolution of certain protein complexes requires multiple, simultaneous mutations, and that this implies they could not have formed by cumulative selection. As a case study, he considers the evolution of *Plasmodium falciparum*, which is the parasite that causes malaria. Specifically, he considers the process through which the parasite developed resistance to the drug choloroquine. He writes:

> The mutant PfCRTs [*P. falciparum* chloroquine resistance traits] exhibit a range of changes, affecting as few as four amino acids to as many as eight. However, the same two amino acid changes are almost always present – one switch at position number 76 and another at position 220. The other mutations in the protein differ from each other, with one group of mutations common to

chloroquine-resistant parasites from South America, and a second clustering of mutations appearing in malaria from Asia and Africa....

Since two particular amino acid changes occur in almost all of these cases, they both seem to be required for the primary activity by which the protein confers resistance. The other mutations apparently "compensate" for side effects cause by these two primary mutations.   *(Behe 2007, 49–50)*

Relying on certain empirical studies, he concludes that the probability of the malaria parasite developing chloroquine resistance by accruing these precise mutations is 1 in 10^{20}. Bacterial populations are so enormous, however, that this small probability is not a barrier to developing resistance. Behe now writes:

Species in which there are fewer living organisms than malaria (again, other things being equal) will take proportionately longer to develop a cluster of mutations of the complexity of malaria's resistance to chloroquine. Let's dub mutation clusters of that degree of complexity – 1 in 10^{20} – "chloroquine complexity clusters," or CCCs. Obviously, since malaria is a microbe, its population is far more vast than any species of animal or plant we can see with the unaided eye. Virtually any nonmicroscopic species would take longer–perhaps much, much longer–to develop a CCC than the few years in which malaria managed it, or the few decades it took for that mutation to spread widely.
                            (Behe 2007, 60)

He is now ready to present his main conclusion:

Recall that the odds against getting two necessary, independent mutations are the multiplied odds for getting each mutation individually. What if a problem arose during the course of life on earth that required a cluster of mutations that was twice as complex as a CCC? (Let's call it a double CCC.) For example, what if instead of the several amino acid changes needed for chloroquine resistance, twice that number were needed? In that case the odds would be that for a CCC times itself. Instead of 10^{20} cells to solve the evolutionary problem, we would need 10^{40} cells. ... If that number has been the same over the course of history there would have been slightly *fewer* than 10^{40} cells, a bit less than we'd expect to need to get a double CCC. ... Put more pointedly, a double CCC is a reasonable first place to draw a tentative line marking the edge of evolution for all life on earth. We would not expect such an event to happen in all of the organisms that have ever lived over the entire history of life on this planet. So if we do find features of life that would have required a double CCC or more, then we can infer that they likely did not arise by a Darwinian process. *(Behe 2007, 63)*

There is some frustration involved in following Behe's argument, since he never actually defines what is meant by the "complexity" of a mutation. However, an analogy will help clarify what he probably has in mind.

Most dogs can be taught to obey simple commands. Some dogs even have sufficient intelligence to learn large numbers of commands. However, it is doubtful that any dog can be taught to respond properly to novel commands presented as sentences written on a piece of paper. Reading comprehension is probably just too complex for canine intelligence. In Behe's terms, there is an "edge" to the level of difficulty a dog can handle. On one side of the edge we find problems that most dogs can solve, while on the other we find problems that are just fundamentally beyond what dogs can do.

Continuing with this, we imagine an experiment where large numbers of dogs are presented with a series of problems that become gradually more difficult. We eventually find a problem that is not quite out of reach for dogs, but which only a minuscule fraction of dogs can ever learn to conquer. We suspect that any problem even slightly more difficult than this will just be fundamentally out of reach. This gives at least a good approximation to the edge of canine intelligence.

Likewise, life poses a series of problems to organisms. Some of these problems are readily solved by evolution while others are just too difficult. For example, everyone would agree that if just a single, simple mutation is needed to improve the fitness of an organism, then evolution is likely to find it. On the other hand, if the entire genome needed to be reorganized in a small number of generations, then we are probably beyond what evolution can do. Somewhere in between we find Behe's edge of evolution.

We now return to the specific case of malaria, continuing to present things from Behe's point of view. The use of chloroquine by humans posed a challenge to the parasite that causes malaria. Would it be able to evolve resistance to the drug? Empirical results suggest that two specific mutations must both occur in the same organism to achieve strong resistance. This was barely within the abilities of what evolution could accomplish, but only because microorganisms have very large population sizes and very short generation times. One suspects that an evolutionary problem even slightly more difficult than evolving chloroquine resistance, especially for animals with relatively small populations and long generation times, would just be fundamentally beyond what evolution could do.

When Behe refers to a "double CCC," he means a problem significantly more difficult than evolving chloroquine resistance. He then argues that life routinely faces such problems, and therefore that something more than standard evolutionary mechanisms are required. That concludes the argument.

We can succinctly reconstruct Behe's logic as follows:

1. Malaria requires two simultaneous mutations to achieve strong chloroquine resistance.
2. Empirical studies suggest the probability of a malaria parasite having both mutations is 1 in 10^{20}.
3. Therefore, a cluster of mutations twice as complex as chloroquine resistance would have probability 1 in 10^{40}.

4. There have been fewer than 10^{40} cells in the history of life on earth.
5. Therefore, Darwinian evolution cannot account for the presence of such mutational clusters in nature, of which there are many.

Now recall the two challenges we posed at the end of Section 5.5. We argued that any probability-based, anti-evolution argument must explain why it is acceptable to reduce probability to combinatorics, and it must explain why the small number at the end has any significance. Behe's answers would seem to be the following:

- The need for multiple, simultaneous mutations rules out any role for cumulative selection, and that is why we can reduce probability to combinatorics.
- The need for specific mutations implies that we cannot dismiss the small probability on the grounds that *something* had to happen. It is like the difference between drawing any two cards at random from a deck versus drawing two specific cards.

How should we reply to all this?

The transition from item two to item three is vulnerable on two fronts. The first is that it uses a premise based on empirical studies of chloroquine resistance in malaria parasites to draw a conclusion about what is possible in any context in any organism. That is, Behe treats the empirical data he cites as though they tell us something about mutations, when a more judicious conclusion is that they are telling us something about malaria.

A related problem comes when Behe starts multiplying probabilities. If we know the probabilities of two independent events individually, then the product represents the probability of those specific events happening in tandem. In this case, the underlying events are two, distinct, given clusters of mutations, each of which has a probability of $1/10^{20}$. Biologist Kenneth Miller explained the importance of this point in his review of Behe's book:

> Behe, incredibly, thinks he has determined the odds of a mutation "of the same complexity" occurring in the human line. He hasn't. What he has actually done is to determine the odds of these two

exact mutations occurring at precisely the same position in exactly the same gene in a single individual. He then leads his unsuspecting readers to believe that this spurious calculation is a hard and fast statistical barrier to the accumulation of enough variation to drive darwinian evolution.   *(Miller 2007, 1055)*

In the language of our dog analogy, Behe's argument is tantamount to carrying out a single experiment on poodles and then drawing a conclusion about what any dog can do in any scenario.

These are salient points, and they ought to make us skeptical as to whether Behe's calculation really amounts to anything. However, the really fatal problem with the argument lies elsewhere. The main problem is that the first bullet point above is not a sufficient justification for reducing probability to combinatorics.

To see why, note that Behe's argument requires that we be able to identify "double CCCs" in organisms. We can take this to mean that we must be certain that multiple, simultaneous mutations are necessary for the evolution of that protein complex relative to some ancestral state. This is essential to Behe's argument since we must be able to rule out any possibility of cumulative selection. Making such a determination entails a thorough understanding of the geometric structure of genotype space in the neighborhood of the protein complex. We therefore find ourselves once more confronting a familiar problem: how can we ever be confident that we know enough about this structure to make such a strong statement?

In the context of chloroquine resistance in malarial parasites, Behe bases his conclusion solely on certain empirical data. He notes that two specific mutations are nearly always present in strongly resistant parasites and concludes that both of these mutations are necessary before the parasite can benefit from either one individually. This claim has been strongly challenged by critics of Behe's work, who note that research has shown his conclusion to be false. Again, Kenneth Miller makes the salient point:

Behe obtains his probabilities by considering each mutation as an independent event, ruling out any role for cumulative selection, and requiring evolution to achieve an exact, predetermined result. Not only are each of these conditions unrealistic, but they do not apply even in the case of his chosen example. First, he overlooks the existence of chloroquine-resistant strains of malaria lacking one of the mutations he claims to be essential (at position 220). This matters, because it shows that there are several mutational routes to effective drug resistance. Second, and more importantly, Behe waves away evidence suggesting that chloroquine resistance may be the result of sequential, not simultaneous mutations, boosted by the so-called ARMD (accelerated resistance to multiple drugs) phenotype, which is itself drug induced.

(Miller 2007, 1055)

Furthermore, the 10^{20} figure is highly dubious as an estimate of the frequency of chloroquine resistance. Biologist Nicholas Matzke writes:

Behe obtains the crucial 10^{20} number from an offhand estimate in the literature that considered only the few CQR [chloroquine resistance] alleles that have been detected because they have taken over regional populations. What is needed, however, is an estimate of how often any weak-but-selectable CQR originates. A study conducted in an area where CQR is actively evolving showed that high-level CQR is more complex than just two substitutions but that it is preceded by CQR alleles having fewer substitutions; moreover, Behe's two mutations do not always co-occur. As a result, CQR is both more complex and vastly more probable than Behe thinks. *(Matzke 2007, 566)*

Our conclusion is that Behe cannot justify the statements he needs regarding the geometric structure of genotype space. His attempt to assign relevant numbers and carry out probability calculations have no importance since the assumptions underlying those calculations have been shown to be false. In the end, Behe's pretensions to mathematical precision amount to very little. His argument is really just that evolution will not move a population from point A to point B if multiple, simultaneous mutations are required. No one disagrees with this, but in practice there is no way of showing that multiple, simultaneous mutations are actually required.

This brings us to the end of our discussion of anti-evolution probability. We have considered several variations on the same basic theme, but they all ultimately ran afoul of the problem we identified back in Section 3.4. A mathematical argument against evolution requires a detailed knowledge of both the probabilistic and geometric structures of protein space (or possibly genotype space, depending on the context). This knowledge is always lacking in practical situations. When you see a probability calculation in some piece of anti-evolution writing, you can be certain that it is based on biologically unrealistic assumptions.

5.11   NOTES AND FURTHER READING

My discussion of the Hardy–Weinberg law was slightly oversimplified, in that I did not discuss what happens in certain extreme cases, such as when all members of the starting population are of types AA and BB. In such cases, it might require two generations for the three genotypes to settle down to their Hardy–Weinberg frequencies. This detail had no bearing on our discussion, and I felt it was more trouble than it was worth to discuss it. For a more detailed treatment, as well as for a very lucid (but technical) introduction to the field of population genetics, I recommend the textbook by Gillespie (2004).

We noted that Dembski's writing about complex, specified information was meant as a general tool for detecting design in the causal history of some event or object. For our purposes, it was only necessary to discuss the proposed application of this tool to biology. However, many other writers have been strongly critical of the whole framework Dembski proposes. For trenchant criticisms in this vein, I recommend the papers by Fitelson, Stephens, and Sober (1999), Godfrey-Smith (2001), and Elsberry and Shallit (2011).

For a general discussion of the role of probability arguments in ID discourse, have a look at the article by Sober (2002). I would also recommend the two articles by Olofsson (2008, 2013).

Anti-evolutionists sometimes direct the BAI toward the origin of life, as
opposed to the origin of novel proteins or genes. The article by Carrier
(2004) provides a thorough refutation of such arguments.

The concept of "specified complexity" introduced in Section 5.6 has
occasionally been floated by scientists. Its first appearance seems to
have been in a book by Leslie Orgel (1973):

> It is possible to make a more fundamental distinction between
> living and nonliving things by examining their *molecular* structure
> and *molecular* behavior. In brief, living organisms are distinguished
> by their *specified* complexity.   *(Orgel 1973, 189)*

Biologist Richard Dawkins and physicist Paul Davies have also
used the concept in their writing. These authors were all using
"specified complexity" in a track one sense. As a casual way of
saying that living things are not just complex, but also embody
independently-specifiable patterns, there is nothing wrong with the
concept. However, the ID proponents claim to have developed a
mathematically rigorous form of the concept, that this work
constitutes a genuine contribution to science, and that they can use
their work to prove that organisms are the result of intelligent design.
It is these claims that are problematic, to put it politely, for reasons
we have already discussed.

For a readable discussion of the "lady tasting tea" experiment, as well as
for discussions of other important moments in the history of
statistics, try the book by Salsburg (2001).

In a paper published a few years after the book *No Free Lunch*, Dembski
elaborated on his approach to specification. There is nothing in this
paper that mitigates the force of the issues I raised in Section 5.7.
However, he does suggest, almost in passing, "bidirectional rotary
motor-driven propeller" as a possible specification of the flagellum
(Dembski 2005, 18). Since we would only come up with this phrase by
looking at the flagellum itself, it is plainly not a detachable
specification in the sense required by his framework. In other words,
this phrase is just a rough description of how the modern flagellum is
seen to operate, as opposed to a pattern that can be described
independently from any knowledge of how modern bacteria move
around.

With regard to Behe's *The Edge of Evolution*, I quoted the reviews by Kenneth Miller (2007) and Nicholas Matzke (2007). A third review by biologist Sean Carroll (2007) also raises many salient problems with Behe's calculations. In particular, all three reviewers point to concrete instances where the sort of protein evolution Behe claims to be impossible is actually seen to occur. Behe (2020), published by the Discovery Institute (an advocacy group that promotes intelligent design), is an anthology of Behe's responses to his various critics, though I am sure you will be unsurprised to learn that I do not believe he has responded effectively. For me the main point is that the research on protein evolution cited by people like Carroll, Matzke, and Miller show how difficult it is to justify a statement that protein space is structured in a way that makes cumulative selection impossible.

An online posting by biochemist Laurence Moran (2020) gathers together his own replies to Behe's book, and provides numerous links to other discussions related to it. Moran emphasizes a further difficulty with Behe's argument that I did not emphasize in my discussion. Specifically, Moran stresses the importance of neutral or nearly-neutral mutations in evolution. The point is that the intermediate stages on the way to some modern protein complex could well have been neutral, as opposed to beneficial, and this dramatically increases the number of mutational routes to the modern structure. Moran's points are well taken. They further emphasize the sheer impossibility of comprehending the geometric and probabilistic structures of protein space in sufficient detail to make mathematical pronouncements about what is possible and what is not.

6 Information and Combinatorial Search

6.1 WHAT IS INFORMATION?

The anti-evolution arguments of the previous chapter asserted that evolution could not account for the complexity of biological adaptations, and they used probability theory to bolster that conclusion. Those arguments failed because we never have enough information to carry out meaningful probability calculations of the sort the anti-evolutionists require.

The arguments in this chapter take a different approach. They focus not on the complexity of adaptations, but instead on the genetic instructions leading to their assembly. These instructions constitute "information," it is said, and neither evolution nor any other naturalistic theory can account for it because natural forces only degrade information. It is claimed that where novel information is found, it must have originated ultimately from an intelligent source. Unsurprisingly, the branch of mathematics used to bolster this argument is "information theory."

Following our previous discussions, we will start with track one considerations about information before taking a brief look at track two. As always, it will not be a problem to skim over the track two details. I will need to make a few statements about exponents and logarithms that may be unfamiliar to you, or which might bring back bad memories from a high school math class. These details provide some useful context and background, but they are not necessary for following the flow of the argument.

Now, "information" is a hard concept to pin down since it has different meanings in different contexts. In everyday conversation it refers to a meaningful message. You ask your colleague in the next office what time the meeting is. He replies, "It's at 2:00." Information has been sent from a sender, your colleague, to a receiver, yourself. Understood in this sense, "information" is a hopelessly subjective concept. For example, you might look at a page written in a foreign language and say, "These words mean nothing to me." The words do not become information until there is a receiver capable of interpreting them.

In light of all this subjectivity, how can we develop a mathematical treatment of information? Mathematical analysis requires precision and objectivity, both of which seem to be lacking in our everyday understanding of the term.

The answer is that even in the everyday sense there is an aspect of information that *can* be quantified. We all have a sense that we learn more from unexpected statements than we do from mundane statements. If a child is offered a free choice between candy and broccoli, it is not very informative to learn they chose the candy. It is much more informative to learn that they chose the broccoli.

Apparently, surprising statements are more informative than mundane statements, and that is what we try to quantify when we pass to track two. How do we assign a number to how surprised we are by a particular piece of information?

Probability theory seems like a useful tool here. A statement is surprising precisely when it has a low probability of being uttered, and a statement is mundane when it has a high probability. Mathematically speaking, we might then say that information content is something possessed by an event in a probability space.

This is progress, but how do we quantify information? To answer that question, we first ask what properties we want our measure to have. The first property we have already discussed: Low probability should correspond to high information and vice versa. A second criterion is that we probably do not want to speak of negative information. While misinformation is a serious societal problem, especially in the political realm, we should probably ignore it for the purposes of our mathematical theory. Thus, our information measure should never take on a negative value. Zero information content is the smallest amount there is.

There is one more property that seems natural: The amount of information you get from two independent events should be the sum of the information content from each one individually. In other words, imagine that you learn a certain amount of information from one event, and then later you learn another amount of information from a completely separate event. Our measure should be such that the amount of information you learn from both events combined ought to be the sum of the two separate amounts.

Summarizing, we want our information measure to have these three properties:

1. Information content should never be negative.
2. Low probability should correspond to high information, and vice versa.
3. The information content from two separate events should be the sum of the information content of each event separately.

A mathematician will quickly notice that this list perfectly describes the logarithm function (with the proviso that we multiply by -1, for reasons that will become clear momentarily). This leads us to the following definition: Suppose that E is an event in a probability space. Let $I(E)$ denote the information content of E, and let $P(E)$ denote the probability of E. Then we define $I(E)$ by the following equation:

$$I(E) = -\log_2(P(E)).$$

We measure information in "bits" (in the same sense that we might measure length in inches or volume in gallons).

No doubt you learned about logarithms in a high school mathematics class, but let me refresh your memory about how they work. The expression "$\log_2(x) = y$" is precisely equivalent to the statement "$2^y = x$." In other words, to evaluate a base 2 logarithm, you ask yourself, "2 to what power will give me x?" For example, we have that $\log_2 4 = 2$ because $2^2 = 4$. We also have that $\log_2 32 = 5$ because $2^5 = 32$.

Let us see how this relates to our definition of information. First, suppose that E is certain to occur, meaning that it has probability 1. We would then compute that

$$I(E) = -\log_2(P(E)) = -\log_2(1) = 0,$$

where the last part of the equation follows from the fact that $2^0 = 1$. This makes sense. If an event is certain to occur then no information is learned when it actually occurs.

Now suppose we flip a fair coin. How much information do we learn when we find out the coin landed heads? In this case, there are two equally likely outcomes. If we let H denote the event of getting heads, then we compute:

$$I(H) = -\log_2(P(H)) = -\log_2\left(\frac{1}{2}\right) = 1,$$

where this time the last part of the equation follows from the fact that $2^{-1} = 1/2$. Since probabilities are fractions, their logarithms are negative, and this is why we had to multiply the logarithm by -1. (Recall that we want information content to be positive.)

Thus, you learn one bit of information when you find out the coin landed heads. More generally, you get one bit of information from learning the result of any 50/50 experiment.

Finally, what happens if E is impossible, meaning it has probability 0? In this case we would compute:

$$I(E) = -\log_2(P(E)) = -\log_2(0).$$

Now we have a problem. If we ask "2 to what power gives us 0?" then the answer is that there is no such power. You cannot raise 2 to a power and end up with 0. Therefore, it would seem that the information content is undefined in this case. This, too, fits well in our scheme since it implies there is no possibility of learning any information from an event that cannot possibly occur.

These examples show that our logarithmic measure of information works rather well. What we have seen to this point are the first steps toward a useful mathematical theory of information.

The first to reason along these lines was Claude Shannon, who pioneered the study of information in a famous 1948 paper called "A Mathematical Theory of Communication" (Shannon 1948). He was studying certain practical problems regarding the efficient transmission of information over a channel such as a telegraph wire. With further development, his approach allows you to answer questions like, "What is the maximum amount of information that can be transmitted across this channel?" or "How can I compress this information for transmission without losing anything?"

However, like all great ideas, Shannon's work quickly found applications in areas far removed from anything he considered. For example, Shannon published his ideas at roughly the same time that the first mechanical computers were being built. With the advent of computer science as a serious discipline, questions about the efficient manipulation and storage of information took on considerable importance.

Another domain of application is biology, as we will now discuss.

6.2 INFORMATION THEORY AND BIOLOGY

Information talk is commonplace in biology, and it is not hard to see why. It is difficult to say anything about genes or DNA without in some way using concepts related to information. Genes are said to contain the instructions for making an organism. The cell is said to contain machinery for translating and transcribing the information in DNA. We commonly speak of the genetic code. Different gene sequences that code for the same amino acid or protein are said to be synonymous, implying that they have the same meaning.

In light of our remarks in Section 6.1, this might seem a little strange. Information, both in the everyday sense and in the mathematical sense, is closely tied to the idea of there being a sender, a receiver, and a transmission channel. Is there anything playing those roles in the biological context? Moreover, information in human affairs seems intimately related to communication, but what is the analog of this in biology? After all, DNA does not know it contains information. At the cellular level, everything just plays out according to the principles of physics and chemistry. Scientists have a pretty good understanding of the physical processes that transform a genotype into a phenotype, and you can understand these processes perfectly well without making any reference to information.

We might then wonder whether information is genuinely a fundamental concept in molecular biology or is instead just a useful source of metaphors. This question has been the cause of considerable controversy among scientists and philosophers for many years. In a 1995 article, biologists Eörs Szathmáry and John Maynard Smith took the former view:

> A central idea in contemporary biology is that of information. Developmental biology can be see as the study of how information in the genome is translated into adult structure, and evolutionary biology of how the information came to be there in the first place. Our excuse for writing an article concerning topics as diverse as the origins of genes, of cells and of language is that all are concerned with the storage and transmission of information.
>
> *(Szathmáry and Maynard Smith 1995, 231)*

Representative of the latter view is philosopher Paul Griffiths:

> It is conventional wisdom that insofar as the traits of an organism are subject to biological explanation, those traits express information coded in the organisms' genes. … I will argue, however, that the only truth reflected in the conventional view is that there is a genetic code by which the sequence of DNA bases in the coding regions of a gene corresponds to the sequence of amino acids in the primary structure of one or more proteins. The rest of "information talk" in biology, and the claim that biology "is, itself, an information technology," is on a par with the claim that the planets compute their orbits around the sun or that the

economy computes an efficient distribution of goods and resources. It is a way to talk about correlation that, in some cases, allows a useful application of the mathematical theory of communication and in others plays no theoretical role but merely reflects the current cultural prominence of information technology. *(Griffiths 2001, 395)*

The proper role of information talk in biology remains a rich source of discussion among biologists and philosophers to this day. Peruse the relevant academic journals, and you will note a steady accumulation of articles on the subject. Scholars occasionally propose new ways of conceiving of information in biology, and these proposals are inevitably criticized by other scholars.

For our purposes, a minimal takeaway from this body of literature is this: If you want to discuss biological information, especially if you are going to argue that there is something about it that utterly confounds a successful, long-standing, scientific theory, then it is essential that you be crystal clear about what you mean by "information." I have emphasized Shannon's view of information because that is the version that is used most commonly. Mathematicians and biologists have also developed other ways of measuring information, and they have their uses in various contexts. For this reason, when discussing questions about the information content of genomes, it is essential to explain precisely what you mean by "information," as well as how you intend to measure it.

The conventional view among scientists is that it is fine to discuss information in the everyday sense, which typically involves notions of meaning and purpose, but then you must abandon any hope of a precise, mathematical treatment. Alternatively, you can devise a precise mathematical treatment, but only if you ignore the aspects of meaning and purpose so important to our everyday understanding.

We shall see that anti-evolutionists routinely elide that distinction, and this makes their arguments entirely unconvincing.

6.3   HOW EVOLUTION INCREASES GENETIC INFORMATION

Aficionados of evolution/creation disputes might recall a certain incident from 1997. In an article written after the fact, biologist Richard Dawkins offered this description of what occurred:

> In September 1997, I allowed an Australian film crew into my house in Oxford without realizing that their purpose was creationist propaganda. In the course of a suspiciously amateurish interview, they issued a truculent challenge to me to 'give an example of a genetic mutation or an evolutionary process which can be seen to increase the information in the genome'. It is the kind of question only a creationist would ask in that way, and it was at this point I tumbled to the fact that I had been duped into granting an interview with creationists – a thing I normally don't do, for good reasons.   *(Dawkins 2003, 91)*

In light of our discussion in Section 6.2, you can see why Dawkins reacted as he did to that question. It is impossible to give a short answer to it of the sort that might be appropriate for an interview in a documentary. Instead, you first have to explain clearly what information is and how you intend to measure it, both of which are tricky questions. How will we know that some process has "increased" genomic information unless we can quantify the amount of information both before and after the process in question?

We will answer the filmmaker's question in this section, but first there are some other issues to address. For instance, we might wonder why anti-evolutionists take any interest in information at all. Why do they think there is something about it that presents a point of attack against evolution?

Part of the answer is that information is fundamentally imma-terial. Information might be expressed in some material form, but it is immaterial nevertheless. The same information can be carved in stone, written in ink on paper, shouted across a room, or stored on a hard drive, but the information itself is independent of its substrate. Ink on a page is governed by the same physical principles regardless of whether it expresses meaningful words or meaningless scribbles, but the former contains information while the latter does not.

That information is immaterial warms the hearts of anti-evolutionists. As they see it, information is a fundamental attribute of living systems, and since it cannot be understood in purely material terms, it forces us to introduce a nonmaterial dimension.

The rest of the answer is that a track one conception of information is all about sending messages between senders and receivers, and such activities are only engaged in by intelligent agents. Shannon's famous paper was called, "A Mathematical Theory of Communication," and not, "A Mathematical Theory of Things Natural Forces do on Their Own." If genes contain information, and if information is about communication between intelligent agents, then this suggests that genetic information in some way represents the will of an intelligent agent.

Consequently, information talk is ubiquitous in the anti-evolutionist literature. They are constantly challenging scientists to explain the origin of novel genetic information. Sometimes they are really nasty about it. For example, here is Phillip Johnson, from his 2000 book *The Wedge of Truth*:

> If the evolutionary scientists were better informed or more
> scientific in their thinking, they would be asking about the origin
> of information. The materialists know this at some level, but they
> suppress their knowledge to protect their assumptions.
>
> *(Johnson 2000, 167)*

This is another of those times where, if you possess any skeptical impulses at all, they should be triggered by the thought of biologists needing a law professor to tell them how to do their jobs. We should also note the rhetorical trickery involved in transforming the evolutionary scientists of the first sentence into the materialists of the second. When carrying out their professional work, evolutionary scientists have no interest in materialism, or in any other metaphysical viewpoint for that matter.

Anti-evolutionists see themselves as great philosophers of information, and they are constantly boasting of their own contributions to its conceptual development. In 2013, they published a large volume called, *Biological Information: New Perspectives*, which was the proceedings of a conference held primarily to showcase the ID view of this subject (Marks et al. 2013). The old perspective, in their telling, is that evolution has little trouble explaining the origin of novel genetic information in terms of known physical mechanisms. As they see it, they have shown mathematically that this view is untenable.

In Section 5.6, we encountered William Dembski's notion of "complex, specified information," by which he just meant improbable things that fit a pattern. You might have wondered why he used the term "information" in his writing. If "probability" was what he meant, then he could simply have said *that* and not have used the term "information" at all. He explained his word choice in his 2004 book *The Design Revolution*. He writes:

> Specified complexity (or complex specified information, as it's also called) is therefore a souped-up form of information. To be sure, specified complexity is consistent with the basic idea behind information, which is the reduction or ruling out of possibilities from a reference class of possibilities. But whereas the traditional understanding of information is unary, conceiving of information as a single reduction of possibilities, specified complexity is a binary form of information. Specified complexity depends on a dual reduction of possibilities, a conceptual reduction (i.e., conceptual information) combined with a physical reduction (i.e., physical information).   *(Dembski 2004, 137–138)*

We have already seen that scientists have good reasons for finding Dembski's framework to be unworkable in practice. Do also note the presumption involved in the first sentence, where we are casually informed that specified complexity is known by more than one name.

Since scientists do not use any of this terminology in the particular, idiosyncratic way that Dembski uses it, he is the only one assigning any names at all to these concepts.

Biologists find this anti-evolutionist obsession with information a little strange. Information theory has played a role in evolutionary biology since at least 1961, when geneticist Motoo Kimura published a famous paper using Shannon's conception to quantify the growth of genetic information in the course of evolution (Kimura 1961). Moreover, it is clear that there is *some* sense in which genetic information has increased in the course of natural history. The earliest life forms had small genomes that coded for simple organisms. Modern life forms have large genomes that code for complex organisms. It seems hard to believe that scientists would have overlooked something so obvious, and, in fact, they have not. The reality is that as soon as a precise definition of "information" is provided, it is never difficult to explain information growth in the course of evolution.

For example, suppose we think of genetic information in Shannon's sense. We could argue that each nucleotide on a string of DNA is chosen from among four possible bases. If we treat these four possibilities as equally likely, then we can say that each base conveys two bits of information (since $\log_2 4 = 2$). This is a slight oversimplification since the four bases are actually not equally likely, but this detail is not important for our argument.

Viewed in this way, the problem of information growth is really the problem of creating new genes. The solution to the problem is found primarily in a well-known process called gene duplication. As the name suggests, it sometimes happens during DNA replication that a stretch of genetic material gets duplicated. Literally, you end up with two copies of a gene where previously you only had one. The two copies can then diverge, with the result being more genes at the end than you started with. This is plainly an increase in genetic information.

There is nothing speculative or cutting edge about this. Gene duplication is a common and well-understood process, and it has been part of the biologist's toolkit for many decades. For example, here is biologist Julian Huxley, writing in 1942:

> The converse is known as *duplication*, when a portion of a chromosome comes to be repeated, occurring twice instead of once, either in the form of a translocation to another chromosome, or of a "repeat" within the same chromosome, often immediately adjacent to its original position. ... They are of much greater ultimate importance, since they constitute the chief method by which the number of genes is increased, thus providing duplicate factors, and the opportunity for slight divergent specialization of homologous genes, giving great delicacy of adjustment.
>
> *(Huxley 1942, 89)*

Modern gene sequencing techniques make it easy to identify when duplications have occurred, and it is commonplace to find them implicated in major evolutionary transitions.

Biologists have been pointing this out to anti-evolutionists for many years now, but to no avail. For example, Phillip Johnson recounts a conversation he had with physicist Paul Davies, in which Davies mentioned gene duplication:

> When I asked Davies about this, his reply gave me the impression that he thinks that natural selection increases genetic information by preserving copies that are made in the reproductive process. I am afraid this misses the point. When two rabbits reproduce there are more rabbits, but there is not any increase in genetic information in the relevant sense. If you need to write out the full text of the encyclopedia and have only page one, you cannot make progress toward your goal by copying page one twenty times.
>
> *(Johnson 2000, 59)*

It is true that the mere duplication of a gene does not increase information in the sense of creating a new functional structure for the organism. The problem, however, is that Johnson's response just ignores half of the process. Duplication by itself increases the information storage capacity of the genome, in the same sense that inserting a blank flash drive increases a computer's information storage capacity. New information is then created when the genes subsequently diverge, just as a computer is storing more information than before once you put a new file onto the drive.

A similar response has been offered by Georgia Purdom, a young-Earth creationist:

> As has been said many times on this site, duplications ... and mutations do not add new information to the genome. Duplications are the result of duplicating existing genetic information, and mutations alter existing genetic information (whether original or duplicated). Neither of them adds new information.

> Think about it this way: if I give someone a copy of a book they already own, then they don't have any new information, just a copy of information they already had. If I subsequently take a marker and mark out some of the letters or words in the copy of the book I gave them, they still don't have any new information – just a messed up copy of one of the books.　　　*(Purdom 2008)*

Again, it is not duplications by themselves, or mutations by themselves, that account for information increase. Rather, it is duplication, followed by mutations in the duplicate copy that increase information. If I give someone a second copy of a book they already own, and then I change the text of the second copy by writing in some words of my own, then the recipient of the second book certainly does have some new information.

Shannon's view of information is useful in many contexts, but as applied to genes it seems to miss something important. If we think in Shannon's terms, then any random sequence of DNA bases contains lots of information, even if that DNA does not code for anything useful. Construed in this way, the quantity of information in a genome is strictly a function of the number of DNA bases, but somehow this does not seem to capture the main issue. What we *really* need is something like a measure of useful information. Modern organisms can do lots of things that ancient organisms could not do. They have acquired new functionalities in the course of natural history. Can we measure and explain this increase in "functional information"?

Whatever such measure you define (see Section 6.11 for one possibility), there is no challenge in explaining how evolution increases functional information. The required mechanism is natural selection acting on random variations, as we have discussed at length elsewhere in the book. New functional information arises in a gradual, stepwise manner. When selection surveys the extant variation in a population, thereby increasing the frequencies of certain genes over others in future generations, the result is an increase in functional genetic information.

There is another sense in which natural selection can be said to increase a genome's information content. We tend to think of genetic information solely in terms of what is needed to construct an organism. That is, "genetic information" means "the information needed to build a phenotype from a genotype." This is fine, but the genome also encodes information about the environments in which organisms find themselves. Evolution causes populations to become better adapted to their environments. As a result, the genomes of recent populations encode more environmental information than those of ancient populations. We could even say that natural selection is an information conduit from the environment into genomes. Since this point will arise naturally in Section 6.8, we will defer further discussion until then.

The point is simply that the only challenge "information" poses to evolution is defining it with sufficient precision to understand what is being asked. By any of the standard definitions used by biologists, there is nothing puzzling about how known mechanisms can explain information growth in evolution. The pretensions of ID proponents to have devised novel understandings of the concept useful for drawing grand biological conclusions have no merit.

This has not stopped anti-evolutionists from serving up a variety of information-based arguments against evolution. We shall consider the most important of these in the remainder of this chapter.

6.4   THE BASIC ARGUMENT FROM INFORMATION

We come now to the basic anti-evolutionist argument in this area, which is the claim that mutations can only degrade information. In other words, evolution is based on the idea of stringing together useful mutations, but this is said to be impossible because mutations can only break things. Here is a standard presentation of the basic argument, from engineering professor Andrew McIntosh:

> At the molecular level, the laws of thermodynamics do not permit step changes in the biochemical machinery set up for a particular function performed by the cells of living organisms. Thus, random mutations always have the effect of increasing the disorder (or what can be defined as logical entropy) of any particular system, and consequently decreasing the information content.
>
> *(McIntosh 2009, 375)*

Claims of this sort are ubiquitous in anti-evolutionist writing. McIntosh has framed his argument in the language of thermodynamics, which is the subject of Chapter 7, but the claim that mutations always degrade information is standard in anti-evolutionary discourse.

Biologists regard this claim as bizarre. It is logically impossible for all mutations to degrade information because mutations can reverse themselves. If the mutation from A to B represents a loss of information with respect to some measure, then it would seem that the reverse mutation from B back to A would have to be a gain of information. Moreover, we have seen that the existence of beneficial mutations is a simple empirical fact, and gene duplication adequately explains how it is possible to increase the size of the genome.

There is another peculiar aspect to McIntosh's argument. The question of whether complex biological adaptations can evolve in a stepwise manner does not seem to have anything at all to do with information or thermodynamics. Instead, it has to do with the structure of the adaptation itself. Either the adaptation can be broken down into small mutational steps or it cannot. Evolutionists say that all adaptations studied to date can be so broken down while anti-evolutionists deny this, but for the moment the only point is that this dispute has nothing at all to do with information or thermodynamics.

At this point we might wonder if McIntosh is using some idiosyncratic notion of how to measure information. Indeed, shortly after the previous quote he presents his own view of how we ought to conceive of information (note that he had previously defined a "machine" to be "a functional device which uses energy"):

> Under this approach we can now go further to define information very specifically. We propose the following definition:
>
> *Definition:* Information is not created by purely material processes and carries (in coded form or otherwise) an intelligence.
>
> But though information is not matter and not energy, it nevertheless *always has an effect on the local thermodynamics of any working system.* We thus propose the following:
>
> *Proposition:* Where information interacts with matter, it always use a machine.   *(McIntosh 2009, 378)*

Vague definitions of this sort are also common in anti-evolutionist literature, which is why I emphasized the need to be crystal clear about your intentions when discussing this topic. A mathematician would protest that not only has McIntosh not defined information "very specifically," but in fact he has not provided any coherent definition at all. He has made an assertion about how information is created, but his proposed definition tells us nothing about what information is or how we measure it. It is likewise unclear what it could mean to say that information, which is said to be created by nonmaterial means, can nonetheless interact with matter.

As we saw in Section 3.2, formal mathematics is usually presented in the form of rigorous definitions followed by propositions or theorems, which are then proved with meticulous logic. McIntosh has imitated some of the form of mathematical writing, but his arguments are fatally imprecise, and he certainly has provided no reason for thinking genetic information cannot increase through the standard mechanisms.

6.5 IS DNA COMPARABLE TO HUMAN LANGUAGE?

Perhaps the problem is not that the genome contains information per se, but rather that it contains symbolically-encoded information. Many human languages employ an alphabet out of which meaningful words are formed. This is in some way analogous to the way triplets of DNA "letters" can be seen as encoding information about proteins. Just as human alphabets and languages are indicative of intelligent design, we might argue by analogy that DNA must likewise be the product of design. One advocate of this view is engineer Werner Gitt, who endorses young-Earth creationism.

In his book *In the Beginning Was Information*, he put forth a series of alleged theorems about information that are meant to prove that evolution cannot account for its growth over time. He lays out the plan of the book like this:

> Since the concept of information is so complex that it cannot be defined in one statement, we will proceed as follows: We will formulate various special theorems which will gradually reveal more information about the "nature" of information until we eventually arrive at a precise definition. *(Gitt 2001, 45)*

Let us consider some representative theorems. He writes (in the quotes that follow, all emphases are in the original):

It should now be clear that *information*, being a fundamental entity, cannot be a property of matter, and its origin cannot be explained in terms of material processes. We therefore formulate the following fundamental theorem:

> **Theorem 1:** The fundamental quantity **information is a nonmaterial** (mental) entity. It is not a property of matter, so that purely material processes are fundamentally precluded as sources of information.
>
> *(Gitt 2001, 47)*

Skipping ahead, we come to this:

> Information is always based on the will of a sender who issues the information. It is a variable quantity depending on intentional conditions. Will itself is also not constant, but can in its turn be influenced by the information received from another sender. Conclusion:

> **Theorem 2:** Information only arises through an intentional, volitional act.
>
> *(Gitt 2001, 48)*

Gitt then presents what he describes as five different aspects of information. He takes for granted that information must be expressed in the form of a code that employs symbols in some way, such as an alphabet or the sequence of base pairs in DNA. He then describes the lowest level of information as "statistics," meaning the actual sequence of symbols comprising the message. Gitt emphasizes that Shannon's theory treats information purely at the statistical level, with no thought to any meaning the symbols might convey. The second level is then "syntax," meaning that certain sequences of symbols become meaningful because of their ordering. The third level is "semantics," which refers to the meaning of the message. He writes:

> Both the sender and the recipient are mainly interested in the meaning; it is the meaning that changes a sequence of symbols into information. ...

Theorem 13: Any piece of information has been transmitted by somebody and is meant for somebody. A sender and a recipient are always involved whenever and wherever information is concerned.

It is only at the semantic level that we really have meaningful information, thus we may establish the following theorem:

Theorem 14: Any entity, to be accepted as information, must entail semantics; it must be meaningful.

(Gitt 2001, 69–70)

For completeness, I will mention that Gitt's final two levels are "pragmatics," meaning roughly the effect the information produces in the recipient, and "apobetics," meaning roughly the intent of the sender in formulating the information in the first place. Gitt goes on for many pages about these five levels, pausing occasionally to lay out his theorems.

He eventually builds up to his promised definition of "information":

The domain A of definition of information includes only systems which encode and represent an abstract description of some object or idea …. This definition is valid in the case of the given examples (book, newspaper, computer programs, DNA molecule, or hieroglyphics), which means that these lie inside the described domain. When a reality is observed directly, this substitutionary and abstract function is absent, and examples like a star, a house, a tree, or a snowflake do not belong to our definition of information …. The proposed theorems are as valid as natural laws inside the domain we have just defined. *(Gitt 2001, 85)*

Later in the book he presents his grand conclusion: "In the light of the information theorems all materialistic evolution models are useless and are thus rejected." (Gitt 2001, 136). This last was presented in bold-face type.

As is clear even from the few quotations I have presented, Gitt's presentation is full of the swagger and bravado typical of creationist discourse. However, it is hard to discern an actual argument in his writing. Though he asserts, vigorously, that information can only come from an intelligent source, he presents no argument for that conclusion.

A mathematician reading his book would marvel at its unusual structure. "Theorems" are generally understood to be precise statements of logical equivalence. That is, they assert that certain assumptions imply certain conclusions as a matter of logic. As we discussed in Section 3.2, in real mathematics, the thing that appears next after the statement of a theorem is a careful, deductive proof establishing its correctness.

Gitt's theorems are not of that sort. For him, a theorem is just an assertion he believes to be correct, and he seems to have little interest in expressing himself with careful, precise prose. He could have presented his beliefs simply as a series of bullet points. Calling them theorems is just a cynical way of aping mathematical language without having to do the hard work of mathematical analysis.

Moreover, in real mathematics, definitions precede theorems. You must first have a precise understanding of what the objects are before you can prove theorems about them. Presenting a series of theorems from which a definition is supposed to emerge is just weird. Upon finally arriving at Gitt's definition, a mathematician is likely to feel cheated. We were promised a definition of "information," but we get instead only a vague account of the domain in which the concept of information is claimed to apply. At no point in his book does Gitt answer questions about how he proposes to measure information content, which is a serious omission considering his frequent assertions that evolution cannot increase information.

Shorn of any pretense of mathematical seriousness, Gitt is just making a crude analogy between DNA on the one hand and written communication among people on the other. Just as humans exercise conscious intent in formulating and disseminating information, he asserts, so too must DNA have arisen in the mind of a designer. That is the entirety of his argument.

We should be very skeptical of this analogy, especially in light of our discussion at the end of Section 6.1. It is one thing to use information talk in casual discussions of DNA, but it is quite another to use that metaphor as the basis for grand pronouncements about what is possible and what is not.

As we have noted, this frustrating lack of precision regarding the definition of information is very common in anti-evolution writing. In that regard, we should take note of a more recent paper by Gitt, this time coauthored with Robert Compton and Jorge Fernandez. Gitt was a participant in the ID conference *Biological Information: New Perspectives* mentioned in Section 6.3. Gitt, Compton, and Fernandez essentially reprise the arguments that we have discussed from Gitt's book. They are explicit that information must have four attributes: a code plus syntax, meaning, expected action, and intended purpose. (By "expected action" they mean roughly that information conveys a desire on the part of the sender that the receiver undertake some action.) They then write:

> All four attributes described above are necessary to
> unambiguously distinguish this subset (category) of information.
> Due to this, the formal definition of Universal Information (UI)
> stated below incorporates all four of these distinguishing
> attributes.
>
> **A symbolically encoded, abstractly represented message
> conveying the expected action and the intended purpose.**
>
> *(Gitt, Compton, and Fernandez 2011, 16)*

They go on to assert that "Universal Information" is always produced by intelligent design, and that organismal genomes comprise an instance of it.

However, we have already noted that DNA does not know that it contains information. At the cellular level, everything is just physics and chemistry. So it is very hard to see what "expected action" and "intended purpose" could possibly mean in this context. When DNA is in the presence of the correct cellular transcription systems certain effects are seen to occur, but it is just an abuse of language to describe this in intentional terms. Gitt, Compton, and Fernandez have built notions of intelligent design right into their definition of information. They are welcome to do so, but then it is very unclear that DNA actually contains information at all.

I have belabored this discussion of Gitt's writing for two reasons. The first is that he was the creationist speaker I mentioned all the way back in Section 1.3, the one who received a standing ovation and was praised for having presented an especially powerful apologetic argument. The argument in his conference presentation was precisely the one described here. Gitt is highly regarded among young-Earth creationists, which lends some importance to his writing.

The second reason is that his willingness to hold forth about information, while never stating anything with sufficient precision to make cogent statements about biology, is entirely typical of anti-evolution writing in this area. As I noted in Section 1.3, where anti-evolutionists see a devastating argument against evolution, scientists just see an absurd caricature of a major branch of mathematics.

6.6   PROTEIN SPACE REVISITED

Let us return to the anti-evolutionist's search metaphor. For our present purposes, we will assume it is "protein space" that needs to be searched.

Eden's argument at the Wistar conference, and the probabilistic anti-evolution arguments discussed in Chapter 5, focused primarily on the size of the space. It was argued that the rarity of functional proteins in the vast space of possibilities militated against the ability of evolution to find anything useful. We replied that any argument of this sort must pay careful attention to both the probabilistic and geometric structure of the space. If most of protein space has a probability close to zero of ever appearing in nature, then the vastness of the space as a whole is irrelevant. And if the functional proteins are arranged like stepping-stones, then it does not matter that they are rare in the space overall. In such a case, the very rarity of functional proteins could actually make them easier for evolution to find since deviations from the "one true path" will be quickly punished by natural selection.

A recent line of argument in anti-evolutionist writing takes aim specifically at the geometric structure of the space, rather than at the probabilistic structure. It is claimed that useful proteins are actually not arranged like stepping-stones. Instead, the argument continues, they represent tiny, isolated, islands of functionality in an ocean of useless proteins.

Their main piece of evidence in defense of this claim is a series of "mutagenesis" experiments carried out by protein chemist Douglas Axe, the results of which were published in Axe (2004). Such experiments involve inducing mutations in specific genes for the purpose of better understanding either the gene's function or the protein it produces. Experiments of this sort have been part of biology's toolkit since the 1920s, when it was discovered that X-rays could induce mutations. Whereas X-rays and other natural mutagens induce random mutations, today the technology exists to induce mutations at specific sites along a gene's sequence. Appropriately, this technique is known as "site-directed mutagenesis."

Axe studied an enzyme called TEM-1 penicillinase, which breaks down penicillin and other antibiotics. The idea was to produce a large library of mutants and to determine the fraction of them that retained some function, even if just in a rudimentary form. In the terminology of our search metaphor, Axe sought to investigate the neighborhood in protein space surrounding this particular enzyme. The conclusion of the experiment might, in principle, have been that a large fraction of the mutants retained at least some function, implying that this neighborhood was teeming with possible stepping-stones for evolution to exploit. However, the actual finding was different. Axe found that functional mutants were exceedingly rare, implying that this particular enzyme was exquisitely sensitive to any change.

ID proponent Stephen Meyer explains what he takes to be the significance of this result:

> [Axe's] experiments revealed that, for every DNA sequence that generates a short *functional* protein fold of just 150 amino acids in length, there are ten to the seventy-seventh power *non*functional combinations – ten to the seventy-seventh amino acid arrangements – that will *not* fold into a stable three-dimensional protein structure capable of performing a biological function. ... Thus, for every functional gene or protein fold there is a vast, exponentially large number of corresponding nonfunctional sequence through which the evolutionary process would need to search. Axe's experimentally derived estimate placed that ratio – the size of the haystack in relation to the needle – at 10^{77} nonfunctional sequences for every functional gene or protein fold.
>
> (Meyer 2017, 116–117)

To clarify, we should note that functional proteins are not just linear chains of amino acids. Their physical properties cause them to fold into three-dimensional structures, and this structure plays a large role in determining their biological function.

We can distinguish two threads in Meyer's argument:

- A mathematical claim that Axe's experiment permits a strong conclusion to be drawn about the geometrical structure of protein space.
- A biochemical claim that useful proteins are tiny islands of functionality adrift in a sea of uselessness.

Let us consider each of these claims, starting with the mathematical point.

Even taking Axe's results at face value, he plainly did not show that "every functional gene or protein" is surrounded by a vast ocean of nonfunctional sequences. Axe carried out mutagenesis experiments on one part of one specific protein. Since there are some fifty thousand proteins in the human body alone, it is quite a stretch to draw a general conclusion about protein space from a study of just one part of one of them.

However, Axe's results should not be taken at face value. In an online discussion of Axe's paper, biochemist Arthur Hunt noted that Axe did not actually study the naturally occurring TEM-1 penicillinase. For reasons having to do with experimental necessity, he studied a variant of this enzyme that was more sensitive to mutation. In his online essay, Hunt illustrated his points using the metaphor of a hill in which the size of the hill's base corresponded to the accessibility of functionality – the bigger the base the more accessible the function. Hunt writes:

> In terms of our illustrations, Axe's TEM-1 variant is a tiny "hill" with very steep sides ... Obviously, from these considerations, we can see that assertions that the tiny base of the "hill" ... in any way reflects that of a normal enzyme are not appropriate. On this basis alone, we may conclude that the claims of ID proponents vis-a-vis Axe 2004 are exaggerated and wrong. Axe's numbers tell us about the apparent isolation of the low-activity variant, but reveal little ... about the "isolation" or evolution of TEM-1 penicillinase. (Or any other enzyme, for that matter.)
>
> *(Hunt 2007)*

Hunt goes on to note that other experimental approaches have produced much higher estimates on the frequency of functional proteins.

A further point is that Axe's estimate was based on the specific mutations he generated. These mutations were hardly a complete census of all the ways in which his enzyme might have been mutated. Ironically, this is one place where the sheer size of protein space *is* relevant, albeit not in a way ID proponents might like. No mutagenesis experiment could ever hope to sample more than a tiny, highly localized, portion of protein space. Axe, himself an ID proponent, essentially acknowledges this point right at the start of his paper:

> Although the immense size of sequence space greatly limits the utility of direct experimental exploration, the sparse sampling that is feasible ought to be of use in addressing the most basic question of the overall prevalence of function.   *(Axe 2004, 1295)*

It is one thing to say that such work ought to be of use, but it is quite another to suggest it is sufficiently informative to draw sweeping conclusions regarding the soundness of evolution.

In Section 3.4, I remarked that the anti-evolutionist's search metaphor inevitably fails because we can never hope to understand the probabilistic and geometric structures of the space with sufficient detail to declare that they are inaccessible to evolutionary mechanisms. The considerations of the last few paragraphs help to flesh out why I say that.

To dramatize the point, let us return to our analogy of finding a pizza parlor in the downtown area of a major American city. Meyer's argument, based on Axe's work, is tantamount to looking at the storefront immediately in front of you, noting that it is not a pizza parlor, and then deciding there is no pizza to be had anywhere in the city. It would not be reasonable to base such a sweeping conclusion on the results of such a minimal search.

We now turn to the biochemical question raised by Meyer's argument. Our discussion to this point has emphasized why Axe's experiments do not support the grand conclusions ID proponents wish to draw. But can we also point to evidence that the geometrical structure of protein space is actually amenable to evolutionary exploration?

Indeed we can. Decades of research into protein evolution has made it clear that the space is not at all structured as tiny islands of functionality adrift in an ocean of useless sequences. Protein function is nowhere near as sensitive to change as ID proponents assert.

Molecular biologists typically organize proteins into large groupings called families, which are understood to represent proteins that evolved from a common ancestor. More than sixty thousand such families have been identified. Within a family, proteins can differ at large percentages of their amino acid sequences while maintaining the same fold. Biochemists Evandro Ferrada and Andreas Wagner write:

> [E]ven very distant sequences can have the same fold. If two such sequences have the same common ancestor, they are referred to as members of the same *protein family*. Such unambiguous common ancestry can usually be identified for sequences that differ in up to 60 to 70 percent of their amino acids. Two sequences in the same family can be connected through a series of amino acid changes that traverse a fraction of sequence space while leaving the structure unchanged. When common ancestry can be claimed based on criteria such as common aspects of structure or function, families of proteins are grouped into superfamilies. Superfamilies share a common fold and diverge on average around 70 to 80 percent in sequence space. Sets of superfamilies that share the same three-dimensional arrangement of secondary structure are grouped into the same fold. Amino acid sequences with the same fold can be very different. Based on a systematic comparison of many divergent sequences with shared folds ... such sequences can have more than 95 percent divergence.
>
> *(Ferrada and Wagner 2010, 1)*

Moreover, in some cases evolutionarily accessible paths are known to exist to transition from one fold to another. In a discussion of recent work in this area, biochemist Todd O. Yeates writes:

> Various mechanisms by which proteins might undergo evolutionary transitions from one fold to another have been categorized based on structural comparisons of divergent protein families. In addition, in a few cases significant transitions in structure have been demonstrated following one or a few amino acid mutations in a protein sequence.   *(Yeates 2007, R48)*

And from later in the same article:

> The growing evidence for structural plasticity and intrinsic
> disorder in proteins also supports the notion that many proteins
> are able to sample a range of different three-dimensional
> conformations in the cell. This would critically enhance the
> likelihood of evolving stable new protein folds. ...
>
> Landscape models may also be useful in understanding how
> protein folds are related to the entire space of possible protein
> sequences. Experiments ... on bridge-states between different
> protein folds should help illuminate the nature of that landscape.
> One idea that emerges from such a landscape view is that the
> well-established degeneracy between protein sequence and protein
> structure may be important in making transitions between
> different folds possible. In any given protein fold, the identities of
> some amino acids are important, while a great deal of sequence
> variability is permitted in other positions. This general property
> implies the existence of broad wells in sequence space around any
> given fold. This would be an essential feature for enabling
> evolutionary transitions between different protein folds.
>
> *(Yeates 2007, R49)*

There has also been extensive mathematical modeling of protein evolution, with the consistent finding that explorations of the
space through known mechanisms is entirely feasible. An example
is a 2005 paper by biologist Michael Lynch (note that Lynch was
specifically replying to work by ID proponents Michael Behe and
David Snoke):

> [T]he conclusions derived from the current study are based on a
> model that is quite restrictive with respect to the requirements for
> the establishment of new protein functions, and this very likely
> led to order-of-magnitude underestimates of the rate of origin of
> new gene functions following duplication. Yet, the probabilities of
> neofunctionalization reported here are already much greater than

those suggested by Behe and Snoke. Thus, it is clear that conventional population-genetic principles embedded within a Darwinian framework of descent with modification are fully adequate to explain the origin of complex protein functions.

(Lynch 2005, 2224)

These are just a few items from the research literature, chosen, frankly, because the authors happened to provide easily quotable nuggets that are directly on point relative to the concerns of this section. It would be trivial to produce hundreds, even thousands, more examples, all pointing to the same general conclusion: that the geometric structure of protein space is such as to allow extensive exploration through standard evolutionary mechanisms. Presented opposite this litany of evidence, Axe's experiments do not add up to very much.

Let me close this section by observing that this debate is not entirely academic. Insights gained from the study of molecular evolution have already contributed to advances in medicine. Biologist Stephen Stearns writes:

Information on molecular evolution has contributed decisively to some of the most exciting insights of evolutionary medicine: phage therapy, cancer biology, parasite manipulations of the immune system, human reproductive biology, and pathogen outbreaks. Its contributions are far from exhausted.

(Stearns 2020, 9)

Molecular evolution understood within a broadly Darwinian framework improves people's lives in tangible ways. In contrast, intelligent design theory just sits there and does nothing.

6.7 DAWKINS' WEASEL EXPERIMENTS

The remainder of this chapter involves arguments drawn from a branch of mathematics known as "combinatorial search." If you have never seen this subject before it can seem very abstract. It will be helpful to see some of its basic principles illustrated in the context of a concrete example, and that is what we shall pursue here.

The title of this section is drawn from a demonstration presented by biologist Richard Dawkins in his book *The Blind Watchmaker* (Dawkins 1986). The purpose of the book was to present fundamental ideas of evolutionary biology to a lay audience. The book's early chapters were meant to clear up common misconceptions about the theory, and one of those misconceptions was that evolution has to build complex structures just by random chance.

To dramatize the *actual* manner in which evolution is said to build complex structures, Dawkins carried out two separate experiments. Both started the same way. He programmed a computer to recognize the target phrase "methinks it is like a weasel," which is drawn from *Hamlet*. Note that this involves 28 characters (including spaces), drawn from an alphabet of 27 possibilities (26 letters and a space). In both experiments, he had the computer generate random strings of 28 characters in an attempt to produce the target phrase.

This is where the two experiments diverged. In the first experiment, he simply had the computer keep generating strings at random, one right after the other. This was meant to model the idea of evolution having to produce complex adaptations just by random chance. Of course, the computer never got close to the target phrase by this method. The space of possibilities is enormous, and the experiment was such that each sequence of 28 characters had the same probability as any other of appearing.

The second experiment proceeded differently. As before, the computer started by generating a small number of random sequences. This time, however, the computer now scanned the offerings for any strings which, just by chance, had a vague resemblance to the target phrase. In the run of the experiment described by Dawkins, the winning string was:

WDLMNLT DTJBKWIRZREZLMQCO P

This does not look very promising, but notice that the 'e' in the second grouping is actually in the correct place. The other strings were now discarded, and this string served as the starting point of the next generation. A new set of strings was generated from this one, but in such a way that each letter was given some chance of mutating to any of the other letters, with each letter having an equal probability of appearing. Then these new strings were surveyed to find the ones with the closest resemblance to the target phrase, and the process began anew. Dawkins reports that after ten generations the winning phrase was

MDLDMNLS ITJISWHRZREZ MECS P

A quick scan shows that several letters are now in the correct places. The winners after 20 and 30 generations were

MELDINLS IT ISWPRKE Z WECSEL

METHINGS IT ISWLIKE B WECSEL,

and the complete phrase was attained after 43 generations.

Dawkins' point was that evolution by natural selection is sometimes misunderstood to be acting in the manner of the first experiment, but in reality it is far closer to the second experiment. Evolution builds complex structures by accumulating small improvements, with selection ensuring there is no backsliding while waiting for the next improvement to appear.

Dawkins intended this to clarify a conceptual point about evolution, but we can use it as a straightforward example of a combinatorial search problem. We started with a very large space of possibilities. Using the techniques from Section 5.4, we find that the total number of 28-character strings drawn from an alphabet with 27 possibilities is 28^{27}, (that is, 28 multiplied by itself 27 times), which is a 40-digit number. Within this space, we are searching for a small target, which in this case is the phrase "methinks it is like a weasel."

There is also a ranking on the strings that allows us to say that some are better than others, which in this context means that some strings are closer than others to the target phrase. In combinatorial search problems, this ranking is usually called a "fitness function" for the space. To picture what is happening, imagine the target phrase sitting at the top of a large hill. Just a little bit below the top are all the strings that are one character away from the target. One notch below that are the strings that are two characters away, and lower still are the strings that are three characters away. There is a vast ocean of strings that have no letters in common with the target, and you can picture them as not even being on the hill at all.

The picture we have is that of a single, gently sloping hill sitting in the middle of an otherwise flat plane. This picture is commonly referred to as the "fitness landscape" for the problem.

We now try to search the space for the target phrase. We cannot possibly try every possible string because the space is far too big, even for a computer. So we employ an algorithm, which is basically a strategy for deciding which specific strings to sample. In Dawkins' first experiment, we used an algorithm known as "blind search." In other words, we chose our string at random and hoped for the best. This approach is extremely unlikely to be successful in a large space.

In Dawkins' second experiment, we used what is known as a "hill-climbing" algorithm. We started by choosing strings at random, but just by chance there will inevitably be one that is just barely on the hill. We used that as our starting point for the next round, by throwing off random variations in all directions. Again, just by chance a small number of strings will end up slightly higher up the hill than the previous one, and they form the basis for the next generation. This is a much more strategic way of searching the space. After the first generation, most of the vast space can be ignored since it is so far from our starting point that it will never arise during the experiment.

The point is that if you want to be successful at searching a large space, then you had better be clever about choosing your algorithm.

Dawkins' demonstration was very effective at clearing up a common confusion about evolution. His simulation captured enough of the important aspects of evolution to show that cumulative selection will very quickly achieve what blind search will never achieve at all. However, there is also an important difference between the demonstration and the evolutionary process. This difference, noted by Dawkins in his presentation, is that evolution has no notion of a "target phrase." Evolution is not searching for a pre-set goal. It just sort of meanders around the space. This is an important point, but it is not as important as anti-evolutionists think it is.

We will revisit this issue in Section 6.9.

6.8   THE NO FREE LUNCH THEOREMS

Let us quickly review our progress. We have emphasized the extent to which anti-evolutionists rely on the metaphor of a search in discussing evolution. The arguments we have considered to this point have all related to the arrangement of the points within the search space. One line of attack asserted that functional structures represented points of such low probability that they could never be found by known evolutionary mechanisms. A different line of attack asserted that functional structures were so isolated within the space that evolution could never bridge the gaps between them. We found both lines of attack to be seriously wanting.

However, the search metaphor has two parts. One part is the space itself and the arrangement of points within it. The other part is the algorithm used to search the space, which in evolution is natural selection acting on chance genetic variations. In recent years, anti-evolutionists have focused much of their fire on the algorithm underlying the evolutionary process. In particular, they employ a collection of results known as the "No Free Lunch" (NFL) theorems, which were published by David Wolpert and William Macready (Wolpert and Macready 1997).

To understand the anti-evolutionary argument, we first need to generalize the examples of Section 6.7. Mathematicians and computer scientists frequently confront problems of the following sort: There is a large space of possibilities under consideration. Each point in the space has a fitness associated with it, meaning that there is a ranking that tells us that some points are better for our purposes than others. Our goal is to find a point that maximizes fitness.

If the space of possibilities is small, then we can just test each point individually until we find the one with maximum fitness. This is usually not feasible in practical problems because the space of possibilities is much too large. For such problems, only certain points can be sampled, and this means some algorithm must be employed for deciding which points to check. In other words, there must be some formal procedure for deciding which point to examine next, given some knowledge of what has already been searched. The blind search and hill-climbing approaches of Section 6.7 are two examples of possible algorithms.

As in our discussion of Dawkins' weasel experiments, we now introduce a visual metaphor. We can imagine the space of possibilities arranged as the points of the xy-plane. The fitness of any point in the plane can then be viewed as a number on the z axis above the plane. In this way, we get a three dimensional surface known, again, as a "fitness landscape."

The success of an algorithm will depend on the shape of the landscape it confronts. If the landscape consists of a single hill with a clear maximum point at the top, as in the weasel experiment, then a simple hill-climbing algorithm will work quite well. If instead the surface is very rugged, then the algorithm might get stuck at a local maximum, even though there are better points elsewhere on the surface. This is shown in Figure 6.1.

There are many other search algorithms available, some of them very ingenious and complex. However, it always seems to be the case that any algorithm works well on some fitness landscapes and not so well on others. In practice, algorithms are typically devised with specific search problems in mind, so it is not surprising that they work better on some surfaces than on others. It is something of an art to know which sort of algorithm to use on which problem, and it can be an inconvenience trying to determine the appropriate algorithm for your particular problem. What would be *really* nice is an algorithm that worked well on any fitness landscape. That would certainly simplify matters.

According to the No Free Lunch (NFL) theorems, there is no such algorithm. If an algorithm works well on certain landscapes,

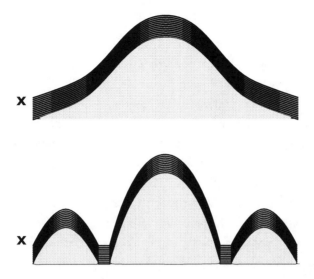

FIGURE 6.1 Two simple fitness landscapes. (Top) A single hill with one
maximum point. Starting from the x, a hill-climbing algorithm will
quickly find the max. (Bottom) Starting from the x, a hill-climbing
algorithm will only get you to the local max at the top of the left-most
hill, but since it cannot go downhill it will never find the global max
point at the top of the middle hill.

then you can be sure there are other landscapes where it performs
poorly. More precisely, the average performance of any algorithm
over all possible landscapes is no better than blind search. Consequently, researchers must tailor their choice of search algorithm to the
problem at hand because there is no all-purpose algorithm to which
they can appeal. Expressed differently, if we think of the targets of
the search as having a high information content, then we can say
the researcher must use prior information about the problem to get
access to the information in the target. You need information to get
information.

We are almost ready to explain why anti-evolutionists believe
there is a point of attack in these observations. It will be helpful,
though, to return once more to our distinction between track one
and track two mathematics. Up to this point I have offered a track
one understanding of what the NFL theorems assert. However, if you

read Wolpert and Macready's paper, you will a find a decidedly track two presentation. The theorems are expressed with copious amounts of jargon and notation, and they will be unreadable to anyone without significant mathematical training. If your intent is merely to understand the main ideas underlying the theorems, then it is fine to remain at a track one level. If instead you presume to use the theorems as the basis for an argument against the fundamental soundness of a successful scientific theory, then you really must engage at a track two level.

With that in mind, let us have a look at what the main NFL theorem *really* says. In keeping with our previous discussions, our point is to emphasize the precision that goes into expressing a proper mathematical theorem. It will not be necessary to parse every symbol, and you are welcome just to skim the following paragraph.

Let us define some notation. We let Ω denote the space to be searched. We let f denote the fitness function, we let Y denote the range of values that f can take on, and we let Φ denote the set of all fitness functions. Let α_i and α_j be two algorithms that search for an optimal point in Ω by searching one point at a time. We imagine that each algorithm has carried out m steps, and that this has produced an ordered m-tuple of measured values of f. We denote this m-tuple by d_m^Y. Finally, let $P(d_m^Y \mid f, m, Y, \alpha)$ denote the conditional probability of obtaining the sample d_m^Y, given f, m, Y, and α. Then we have the following result:

Theorem 6.1 *For any pair of algorithms α_1 and α_2, we have that*

$$\sum_{f \in \Phi} P(d_m^Y \mid f, m, \alpha_1) = \sum_{f \in \Phi} P(d_m^Y \mid f, m, \alpha_2).$$

A serious anti-evolutionary argument based on the NFL theorems would need to engage with all of those letters.

Now let us have a look at how the anti-evolutionists actually use this theorem. William Dembski introduced the NFL theorems into ID discourse in his 2002 book *No Free Lunch*. Roughly, his idea is this: The evolutionary process can be seen as a search of genotype space undertaken using the algorithm of passing random genetic variations through the sieve of natural selection. Evolutionists claim that this algorithm is effective at searching the space, in the sense that it outperforms blind search by finding functional structures and complex adaptations. The No Free Lunch theorems imply that this can only occur if the algorithm has been tailored in some way to the fitness landscape. If evolution can find high-information targets in genotype space, it is argued, it can only be because information was already built into the search algorithm.

Expressed differently, we have seen that when scientists confront specific search problems, they use information about the fitness landscape to tailor their algorithm to the problem. They know the landscape has a certain general shape, and they choose an algorithm known to be effective on landscapes of that shape. Dembski sees this knowledge as information built into the algorithm by the scientists. Analogically, if we hypothesize that Darwinian processes represent successful search strategies, then in some way information has been built into them from the start.

Recall that in Dembski's formalism, complex biological structures represent rare instances of CSI ("complex, specified information," as explained in Section 5.6). If Darwinian mechanisms routinely find such structures in reasonable amounts of time, then they are outperforming blind search. He writes:

> It follows that any success an evolutionary algorithm has in outputting specified complexity must ultimately be referred to the fitness function that the evolutionary algorithm employs in conducting its search. The No Free Lunch theorems dash any hope of generating specified complexity via evolutionary algorithms. *(Dembski 2002, 196)*

He later elaborates:

> The No Free Lunch theorems are essentially bookkeeping results. They keep track of how well evolutionary algorithms do at optimizing fitness functions over a phase space. The fundamental

claim of these theorems is that when averaged across fitness functions, evolutionary algorithms cannot outperform blind search. The significance of these theorems is that if an evolutionary algorithm actually proves successful at locating a complex specified target, the algorithm has to exploit a carefully chosen fitness function. This means that any complex specified information in the target had first to reside in the fitness function.

(Dembski 2002, 212)

At this point, however, we should pause to wonder if the No Free Lunch theorems really apply straightforwardly to evolution. There is certainly a sense in which evolution by natural selection can be said to comprise an algorithm for searching genotype space, but is it a search in the precise, technical sense laid out by the theorems? Can the evolutionary process be formalized as an optimization problem so that all the parameters of the theorem, all those letters we saw in the theorem's formal statement, are assigned proper values?

There is some urgency to this question, since there are at least two obvious ways in which evolution differs from the searches envisioned by the NFL theorems. The first is that the theorems assume that the fitness landscape remains unchanged throughout the search process. This is plainly not the case for evolution. What constitutes fitness in one environment might be entirely different from what constitutes fitness in a different environment. It is sometimes said that biological fitness landscapes are made of rubber, and animals alter a landscape's shape by moving around on it.

The second point of difference is that evolution is not a straightforward optimization process. Philosopher Sahotra Sarkar makes this point very well. After discussing a number of failed attempts to find a parameter that evolution can be said to maximize, he writes,

The conclusion from all these negative results is that, in general, evolution by natural selection cannot plausibly be viewed as a mathematical optimization process leading to the maximization of any parameter that can be interpreted as a fitness.

Consequently, the NFL theorems for optimization are irrelevant
to biological evolution. Dembski's excitement over these
theorems reveals little more than a lack of awareness of
mathematical evolutionary theory. *(Sarkar 2007, 89)*

In light of these considerations, you can understand why David
Wolpert, one of the researchers who first formulated the theorems,
expressed some exasperation with Dembski's argument. In the following excerpt, Wolpert refers back to some remarks about philosophy
made earlier in his review, the details of which are unimportant to
the present discussion:

[D]espite his invoking the NFL theorems, his arguments are
fatally informal and imprecise. Like monographs on any
philosophical topic in the first category, Dembski's is written in
jello. There simply is not enough that is firm in his text, not
sufficient precision of formulation, to allow one to declare
unambiguously 'right' or 'wrong.' ... The values of the factors
arising in the NFL theorems are never properly specified in his
analysis. ... [T]hroughout there is a marked elision of the formal
details of the biological processes under consideration. Perhaps
the most glaring example of this is that neo-Darwinian evolution
of ecosystems does not involve a set of genomes all searching the
same, fixed fitness function, the situation considered by the NFL
theorems. Rather it is a co-evolutionary process. Roughly
speaking, as each genome changes from one generation to the
next, it modifies the surfaces that the other genomes are
searching. And recent results indicate that NFL results do not
hold in co-evolution. *(Wolpert 2002)*

These are salient points, and they certainly suggest that the NFL
theorems have little relevance to assessing questions about evolution.
Many critics of Dembski's writing have raised issues along these
lines, and I will refer you to some of them in Section 6.11. However,

in my view there is an even more serious problem with Dembski's argument, one that has not received sufficient attention.

For the sake of argument, let us assume that the NFL theorems do, indeed, apply to biological evolution. Assessing natural selection's creative abilities requires that we evaluate the efficacy of a particular algorithm acting on a specific problem. NFL addresses only average performance over all possible problems. It therefore offers no reason to believe that selection cannot construct complex adaptations. However, NFL might suggest that selection's ability to ascend the fitness landscapes it actually confronts implies its inability to scale the different landscapes that no doubt exist in some alternate reality. Mutation and recombination, viewed as algorithms for searching genotype space, will be effective only when the landscapes they confront possess certain properties. This makes it reasonable to ask why nature presents us with just the sorts of landscapes that are searched effectively by these mechanisms.

Dembski's answer is that natural selection acts effectively only because CSI was front-loaded into the biosphere. This information is encoded in the fundamental constants of the universe. Physicists have noticed that if you imagine changing the values of certain constants – such as the ratio of the mass of a proton to the mass of an electron or the so-called cosmological constant – in isolation from the others, then the result is a universe that cannot sustain life. This phenomenon is referred to as "fine-tuning." Dembski now writes:

> For starters, [the collection of DNA-based self-replicating cellular organisms] had better be nonempty, and that presupposes raw materials like carbon, hydrogen, and oxygen. Such raw materials, however, presuppose star formation, and star formation in turn presupposes the fine-tuning of cosmological constants. Thus, for f to be the type of fitness function that allows Darwin's theory to flourish presupposes all the anthropic principles and cosmological fine-tuning that lead many physicists to see design in the universe.   (Dembski 2002, 210)

In a later paper coauthored with Robert J. Marks II, Dembski was even more explicit about the intent of his argument:

> We are not here challenging common descent, the claim that all organisms trace their lineage to a universal common ancestor. Nor are we challenging evolutionary gradualism, that organisms have evolved gradually over time. Nor are we even challenging that natural selection may be the principal mechanism by which organisms have evolved. Rather, we are challenging the claim that evolution can create information from scratch where previously it did not exist. The conclusion we are after is that natural selection, even if it is the mechanism by which organisms evolved, achieves its successes by incorporating and using existing information.   (Dembski and Marks 2011, 389)

When Dembski first presented his argument in his 2002 book, many commentators took him to be saying that the NFL theorems in some way imply that evolution cannot construct complex adaptations. It is understandable that they thought so. These theorems were the centerpiece of a book arguing against evolution, and Dembski did, after all, write things like, "The No Free Lunch theorems dash any hope of generating specified complexity via evolutionary algorithms."

The last two quotes make clear that Dembski, and later Dembski and Marks, are really just saying that evolutionary mechanisms achieve such success as they do only because they manipulate information from the environment. In effect, they are saying, "Biologists thought that evolution was able to create eyes, wings, and immune systems from scratch, thereby producing complex, specified information where no such information existed before. In reality, evolution just manipulated previously existing complex, specified information into a new form, and you still have to explain the origin of that prior information."

Understood in this way, this is a *very* strange argument. Most of us did not need difficult mathematical theorems to realize that Darwinian evolution is viable only when nature has certain attributes, and it is not a defect in evolutionary theory that it takes these attributes for granted. It is trivial to imagine alternate realities very much like the one we are in, but in which Darwinian evolution would never have gotten anywhere. The fitness landscapes confronted by evolving organisms arise ultimately from the laws of physics, and therefore the ID argument is tantamount to wondering why those laws are as they are. But determining why the universe has just the properties it does is hardly a problem within biology's domain.

As we suggested in Section 6.3, there is nothing wrong with viewing natural selection as an information conduit between the environment and a population's gene pool. Recall that according to Shannon's conception, information content is something possessed by an event in a probability space. Seen in that way, any physical system that can exist in more than one state can contain information. This is because if the system can exist in more than one state, then there must, in principle, be a probability distribution that describes the likelihood of being in one state versus another. And since the local environments in which gene pools find themselves can certainly exist in many states, it is not an abuse of language to say the environment contains information.

This can actually be an illuminating metaphor. There is a strong sense in which the gene pools of modern organisms can be said to record information about the ancestral environments in which they evolved. The process is not much different from receiving medical information from your doctor and then making lifestyle changes as a result. Just as your doctor gives you information on how to live a healthier life, so too does the environment give information to a gene pool about how better to survive. In the language of information theory, we would say this is communication through a noisy channel because natural selection is not the only mechanism of change, and evolution is not always adaptive. But it is an interesting way of looking at things nevertheless.

Biologists frequently make statements to the effect that evolution can create information. For example, in his 1961 paper referred to in Section 6.3, Motoo Kimura writes:

> We know that the organisms have evolved and through that process complicated organisms have descended from much simpler ones. This means that new genetic information was accumulated in the process of adaptive evolution, determined by natural selection acting on random mutations.
>
> Consequently, natural selection is a mechanism by which new genetic information can be created. Indeed, this is the only mechanism known in natural science which can create it.
>
> *(Kimura 1961, 127)*

Statements of this sort are ubiquitous in the literature of evolutionary biology, but notice that it is specifically genetic information that is intended here. Standard evolutionary mechanisms have the ability both to increase the information storage capacity of the genome through gene duplication, and to refine, via mutation and natural selection, the information stored there, resulting in better-adapted organisms. This is what is meant when scientists attribute to evolution the ability to create new information.

If you now want to play gotcha, and argue that evolution did not *really* create information, but only transformed preexisting information in the environment, then you are welcome to do so. However, it is no great accomplishment to make this observation. If you are just saying that nature has to be a certain way for evolution to work then you can just assert it as obvious without further argument. You do not need to write lengthy books to defend this claim, or to deploy difficult mathematical theorems in support of it.

However, Dembski wants to go further. In his telling, genomes do not merely contain information in some general sense, but instead contain complex, specified information (CSI), which is supposed to be an indicator of intelligent design. He sees his NFL-inspired argument as a rebuke to scientists who claim that evolution can produce both complexity and specification. To the extent that evolution can do this at all, he argues, it is only because the environment must have contained CSI to begin with. He refers to this as "the displacement problem," his idea being that scientists have explained CSI in genomes only by helping themselves to a source of preexisting CSI. He likens this to filling one hole by digging another. We still have to explain the environmental CSI, he argues, and for this purpose we must invoke the action of an intelligent designer.

But Dembski has a lot more work to do if this argument is to be taken seriously. Recall that within his formalism, demonstrating the presence of CSI requires two things: We must carry out a relevant probability calculation to establish complexity, and we must also describe a proper specification in the precise technical sense we saw in Section 5.7. We have already seen that Dembski has no way of carrying out either of these steps for biological adaptations, and he will have no more luck here. Calculating the probability that the fundamental constants would have just the values they do entails knowing the appropriate probability distribution, but we have no useful information for deciding on what distribution that is. Nor do we have any relevant background knowledge that would help us jump through the mathematical hoops required by Dembski's notion of specification. Until Dembski provides these details, he has no basis at all for claiming that the environment contains CSI in his idiosyncratic sense.

We should note that in Dembski's trichotomy of regularity, chance, or design, both regularity and chance are viable options for explaining the origin of the constants. Modern cosmology can tell us almost nothing about how the constants came to be what they are. It could well be that some sort of natural law latent in the origin of the universe determined that the constants had to take on the values we observe, in which case regularity would be the explanation. To illustrate my point, suppose I told a friend that I tossed a coin 100 times and that it came up heads every time. My friend might reply, "That's amazing! 100 consecutive heads is incredibly unlikely." But now I point out that, actually, I used a two-headed coin. What at first seemed incredibly unlikely is seen to be inevitable once all of the relevant information is at hand. I am suggesting that the constants of the universe might have the values they do simply because some unknown physical principle requires that they do. There is nothing in current cosmology to even suggest, let alone imply, otherwise.

Alternatively, it could be that the constants really might have set themselves to a large number of different values, but that ours is just one universe in a vast multiverse, each with its own set of constants, and in this case chance would be the explanation. This scenario can be analogized to winning a lottery. If we only buy one lottery ticket then we are unlikely to win, and this is analogous to finding just the right set of life-supporting constants in a single universe. But if we buy billions of lottery tickets then it is near-certain that at least one will win. This is analogous to finding one life-supporting universe in an enormous collection of such universes.

Dembski needs to eliminate both of these possibilities for his argument to have any force, but he has no plausible way of doing so.

Summing up, the point is that when stripped of all of the irrelevant mathematical formalism, there is nothing more to Dembski's argument beyond wondering why the universe has just the properties it does, and then to assert, based on nothing, that design is the only possibility. The only role for the NFL theorems is to justify the claim that the environment must be a certain way for Darwinian mechanisms to be effective, but that claim is obvious without invoking difficult mathematical formalism. It is an interesting question to ask why the universe is as it is, but it is hardly a problem in biology's domain.

Thus, even taking Dembski's argument at face value there is nothing here to which evolutionary biologists need to pay attention.

6.9 ARTIFICIAL LIFE

The final ID argument we will consider in this chapter unites their thinking about information theory with their rhetoric about combinatorial search. We have one last piece of groundwork to lay before we come to that.

As compelling as the circumstantial evidence for evolution is, it would be better to have direct experimental confirmation. Sadly, that is impossible. We have only the one run of evolution on this planet to study, and most of the really cool stuff happened long ago. If we someday find life on another planet, then we will be able to increase the sample size to two, but that seems unlikely to happen any time soon.

Experimental evolution on microorganisms is possible since they can be cultivated in large populations and reproduce very quickly. Small animals like fruit flies also lend themselves to such experimentation for the same reason, as do plants. Evolutionary experiments on larger animals have been undertaken, but they are limited by a host of obvious practical problems. In short, there has been a lot of illuminating research in the realm of experimental evolution, but by its nature it has little to tell us about the big questions of natural history. For example, you are not going to witness the evolution of a large-scale organ system in the course of these small-scale experiments.

However, Darwinian evolution occurs in any environment in which there is heritable variation, replication, and competition for resources. This suggests the possibility of creating an artificial environment that features those three characteristics. Studying evolution in such an environment could illuminate general principles that arise in any evolutionary environment. Based on such experiments, we might hope to say, "Here's what happened in our synthetic environment, this environment reproduces relevant aspects of the natural environment, and therefore it is plausible to think something similar has happened in natural history."

In particular, we can use computers to create this environment. Our organisms will be computer programs striving to replicate themselves and competing for memory in the computer. We can introduce something equivalent to random mutations, so that as programs replicate themselves there is some possibility of alterations to their code. And we can modify the environment to set up selection pressures in favor of certain abilities the programs might be able to develop. Experiments of this sort have been carried out since the late 1980s, and this area of research is known as "artificial life."

It is commonplace, even in the professional literature, to refer to artificial life experiments as "simulations of evolution." This is convenient terminology for distinguishing these experiments from what happens in nature, and we will use it ourselves, but we should note that it is also slightly misleading. Experiments of this sort are not so much simulations of evolution as they are instances of it. In observing such an experiment you are watching actual evolution take place, albeit in an environment in which the researchers control all of the variables.

An early example of the genre was an experiment carried out by Thomas Ray using a platform he dubbed "Tierra." He started with an 80-line computer program, written in a computer language of his own devising, capable of replicating itself. The only selection pressure was towards efficient replication. Yet even in this incredibly simple environment, astonishing complexity quickly evolved. In a commentary on this experiment, biologist John Maynard Smith described what occurred (note that in the quote, a "satellite virus" is a virus that is completely dependent on its host's machinery in order to replicate):

> The evolutionary behaviour of the system is surprisingly rich. The first important variants to arise are 'parasites', 45 instructions long, which cannot replicate on their own, but use the copying procedure of a neighbour: they are an exact analogue of satellite viruses. Once parasites become common, 'hosts' may evolve immunity; then new types of parasite evolve that are able to attack the new hosts. Evolutionary arms races of this kind have often been postulated and occasionally observed. Next there are 'hyperparasites' which, by an ingenious trick, persuade the parasites to replicate them.   *(Maynard Smith 1992, 772)*

A more recent experiment employed a more sophisticated environment known as "Avida." In results published in 2003, researchers

Richard Lenski, Charles Ofria, Robert Pennock, and Christoph Adami reported that complex functions were seen to originate by cumulative selection, precisely as biologists say happened in nature. From the abstract of their article:

> A long-standing challenge to evolutionary theory has been whether it can explain the origin of complex organismal features. We examined this issue using digital organisms – computer programs that self-replicate, mutate, compete and evolve. Populations of digital organisms often evolved the ability to perform complex logic functions requiring the coordinated execution of many genomic instructions. Complex functions evolved by building on simpler functions that had evolved earlier provided that these were also selectively favored. ... The first genotypes able to perform complex functions differed from their non-performing parents by only one or two mutations, but differed from the ancestor by many mutations that were also crucial to the new functions. In some cases, mutations that were deleterious when they appeared served as stepping-stones in the evolution of complex features. These findings show how complex functions can originate by random mutation and natural selection.
>
> (Lenski, Ofria, Pennock, and Adami 2003, 139)

Experiments of this kind should be seen as a powerful proof of concept for modern evolutionary theory. Biologists theorize that certain processes play out in the course of evolution by natural selection, and now we have concrete, evolutionary systems in which precisely those processes are seen to play out.

Research in this area continues to the present, with ever more complex environments being fashioned to investigate ever more difficult problems. A recent example is work carried out by Yuta Takagi, Diep Nguyen, Tom Wexler, and Aaron Goldman. They used an artificial life experiment to study the selection pressures that might have influenced the relationship between cellularity and metabolism shortly after the origin of life. From their abstract:

> The emergence of cellular organisms occurred sometime between the origin of life and the evolution of the last universal common ancestor and represents one of the major transitions in evolutionary history. Here we describe a series of artificial life simulations that reveal a close relationship between the evolution of cellularity, the evolution of metabolism, and the richness of the environment. *(Takagi, Nguyen, Wexler, and Goldman 2020, 598)*

Artificial life experiments have proven their worth by illuminating processes that are universal to all systems that evolve by Darwinian mechanisms. Researchers have found them to be a useful tool for investigating evolutionary questions that would be hard to study in nature.

Why, then, do the anti-evolutionists think there is something shoddy about the whole enterprise?

6.10 CONSERVATION OF INFORMATION

In their 2017 book *Introduction to Evolutionary Informatics*, Robert Marks II, William Dembski, and Winston Ewert (MDE) continue the story that Dembski began in *No Free Lunch*. The book certainly opens with lofty ambitions:

> In order to establish solid credibility, a science should be backed by mathematics and models. ... The purpose of *evolutionary informatics* is to scrutinize the mathematics and models underlying evolution and the science of design.
>
> *(Marks, Dembski, and Ewert 2017, 1)*

A few paragraphs later, they give a clear statement of their primary thesis:

> Our work was initially motivated by attempts of others to describe Darwinian evolution by computer simulation or mathematical models. The authors of these papers purport that their work relates to biological evolution. We show repeatedly that the proposed models all require inclusion of significant

knowledge about the problem being solved. If a goal of a model is specified in advance, that's not Darwinian evolution; it's intelligent design. So ironically, these models of evolution purported to demonstrate Darwinian evolution necessitate an intelligent designer.   *(Marks, Dembski, and Ewert 2017, 1–2)*

Let us see what this means.

In Section 6.8, we saw that the No Free Lunch theorems imply that there is no universally successful search algorithm. If an algorithm performs well on one problem, then there must be other problems where it performs poorly. As we have discussed, in practice this means that researchers confronted with a specific search problem must tailor their choice of algorithm to fit the problem. That is, the researchers use the information they have about the nature of the problem to choose an effective algorithm.

MDE propose to quantify the information the researcher brings to the problem. Specifically, they refer to this as "active information," claim to provide a rigorous definition of what it is, and then prove theorems about it. Their chief result, which they refer to as a "conservation of information" theorem, then shows that the quantity of information outputted by the algorithm cannot exceed the amount of active information brought to the problem by the researcher.

In responding to this claim, we find ourselves in the same position we were in back in Section 5.6. We saw that Dembski claimed to have produced a piece of mathematical machinery – complex, specified information – that could be used to distinguish things that must have been designed from things that could be explained in some other way. He then proposed to apply this machinery to biology, and claimed thereby to show that intelligent design is in some way implicated in natural history. We decided that we did not need to examine every piece of the machinery, since we only cared about its claimed application to biology.

And so it is here. It would take many pages to explain and analyze the mathematical formalism proposed by MDE, but it would not further our agenda to do so. If they want to apply their techniques to abstract problems in combinatorial search, then they are welcome to do so. Whether such proposed applications have any merit would be a topic for a different book. Our interest here is solely in the manner in which they apply their machinery to biology.

As it happens, and as the earlier quote made clear, they do not propose to apply their conservation of information theorem directly to biology. Instead they target the sorts of artificial life experiments we considered in Section 6.9, as well as other sorts of computer simulations of evolution. Their claim is that all such simulations are unrealistic because the researchers bring active information to their project. For example, with respect to the "Avida" experiment mentioned in Section 6.9, they write:

> [A]vida is a computer program which, its creators say, "show[s] how complex functions can originate by random mutation and natural selection." [C]ontrary to the claims of the authors, the source of the success of Avida is not due to the evolutionary algorithm, but to sources of information embedded in the computer program. A strong contribution to the success of Avida is [sic] stairstep information source embedded in the computer program. (Marks, Dembski, and Ewert 2017, 205)

In our brief discussion of Avida, we mentioned that the programs were able to evolve the ability to carry out complex logic functions. When MDE refer to a "stairstep information function," they mean that the Avida environment rewarded its digital organisms for finding the intermediate steps on the way to the complex function. They later elaborate on why they find this so important:

> The more complex operations are built with simpler operations. A stair step information source to generate more complex operations must signify that the more complex operations are more fit than the simple operations. If this is not the case, the existence of the stair steps is not useful. As the stairs are climbed,

we must be informed we are getting "warmer," i.e. closer to the
result we seek. When a more complex operation degrades into a
simpler one, we are informed that we are getting colder. This
active information source is the reason for Avida's success. As is
always the case, the evolutionary program does not create any
information.   (Marks, Dembski, and Ewert 2017, 209)

It is on the basis of such arguments that MDE dismiss all
computer simulations of evolution as unrealistic, but their logic is
hard to follow. It is self-evident that the Avida organisms found
evolutionary success in part because the researchers created an environment in which success was possible. However, it is equally self-
evident that the algorithm plays a big role in the success. That is,
Avida's organisms achieved success because a particular algorithm
interacted with a particular environment. The algorithm and the
environment are both critical, and therefore it is plainly wrong to say,
"This active information source is the reason for Avida's success."
The fallacy committed by MDE here is essentially the same one we
discussed in Section 5.5.

This point is more easily seen in the context of the weasel experiments described in Section 6.7. We saw two separate experiments
undertaken within the same environment. One employed a blind
search algorithm, while the other employed a hill-climbing algorithm.
The former was unsuccessful, while the latter was successful. The
difference was the algorithm that was used to search the space, and
this is why it is wrong to imply that it was solely the environment
that led to Avida's success.

If MDE want to argue that computer simulations of evolution
are fundamentally unrealistic, then they must show that these simulations are relevantly different from what happens in nature. They
point to two related aspects of Avida that are meant to establish this
disanalogy, but they are not successful in either case.

First, they note that Avida's programmers provided the digital
organisms with a source of information to mine by creating the

environment in the first place. But as we discussed in Section 6.8, nature likewise contains information that populations of organisms can exploit. Natural selection serves as a conduit for transmitting environmental information into the genomes of organisms.

Second, they note that the programmers created a selection gradient that favored intermediate stages on the way to evolving a complex function. But nature also provides selection gradients to organisms. If the intermediate steps all promote the reproductive success of the organisms, then natural selection will favor and preserve those steps. For example, eyes can evolve in a stepwise manner because the intermediate steps represent improvements in visual acuity, thereby improving the reproductive success of their bearers.

Let us elaborate on this last point. Suppose there really were an intelligent designer who established the laws of nature. Further suppose he wants to steer some population of sightless organisms towards the evolution of eyes. In each generation, he surveys all of the offspring for novel variations that improve visual acuity. He allows those creatures to reproduce, and quietly ensures that the others live out their lives without leaving offspring. He does this generation after generation, until, very gradually, a bona fide eye appears. ID proponents would argue that this is not Darwinian evolution, because success was achieved only because an intelligent agent guided the process toward a pre-determined goal.

However, biologists would reply that nature can recapitulate every step of this process without any need for intelligence. Mindless nature, no less than our intelligent agent, can ensure that only the organisms with the best eyesight leave offspring. This is possible because improvements in visual acuity are strongly correlated with an ability to survive and reproduce. What matters is not the presence of intelligence, but only the consistency of the selection mechanism. In our example, it does not matter if our incipient eyes are preserved because an intelligent agent wills it to be so, or if natural selection inadvertently preserves them because they happen to promote increased success in reproduction. Success is achieved in either case because we select for the same thing in every generation.

And so it is with artificial life experiments. Focusing on the fact that intelligent agents set up the environment misses the point. Complexity can be seen to evolve in such experiments because the random mutations introduced into the system interact with the selection gradients in the environment to produce adaptive change. Since the interplay of random variations with selection gradients also takes place in nature, there is no point of disanalogy here between the simulations and the reality.

At this point we might wonder why the ID folks are so relentless in driving home these points. After all, we are once more back to the trivial observation that nature must be a certain way for evolution to work, and we have already noted that no sophisticated mathematics is needed to establish this. So what do Dembski and his coauthors think they have accomplished?

The clearest statement I have found of their intentions comes from a 2011 paper written by Dembski and Marks. They characterize their argument as follows:

> The central issue in the scientific debate over intelligent design and biological evolution can therefore be stated as follows: Is nature complete in the sense of possessing all the resources it needs to bring about the information-rich biological structures we see around us, or does nature also require some contribution of design to bring about those structures? Darwinian naturalism argues that nature is able to create all its own information and is therefore complete. Intelligent design, by contrast, argues that nature is merely able to re-express existing information and is therefore incomplete. (Dembski and Marks 2011, 362)

Biologists will not agree that the central issue is as Dembski and Marks have characterized it. They will also note a strange confla- tion between "biological evolution" in the first sentence and "Dar- winian naturalism" later on. The theory of biological evolution, as understood by scientists, takes no stand at all on questions involving the completeness of nature in the sense Dembski and Marks describe. Instead it is organized around certain strong claims regarding common descent and the ability of natural selection to craft complex structures over time. These are precisely the claims challenged so vigorously in the writing of ID proponents, and they are the claims defended by scientists in their replies. It is these challenges and responses that comprise the debate, such as it is, between scientists and ID proponents, and not any philosophical concerns regarding naturalism or its alternatives.

For all their mathematical jargon and notation, Dembski and Marks have made no serious argument at all. They are merely asking why the universe is as it is. It is a fine thing to ask, and they are certainly welcome to take up this subject with the physicists and philosophers. The fact remains that there is nothing in their theorizing that is relevant to the professional work of biologists, and there is nothing in it to diminish our confidence in evolutionary theory.

6.11   NOTES AND FURTHER READING

The book by Pierce (1961) is an excellent, mostly nontechnical, introduction to the history, development, and applications of information theory. Among many good high-level textbooks on the subject, I especially like the one by Applebaum (1996).

I focused primarily on Shannon's version of information theory since it is both the easiest to understand and the one that is most commonly used in scientific applications. There is another approach to this subject known as "algorithmic information theory," whose main ideas are typically attributed to Kolmogorov, Solomonoff, and Chaitin. The basic idea is to define a measure of complexity based on the length of the shortest computer program that would be needed to produce the piece of information in question. ID proponents sometimes refer to algorithmic information theory in their work, but not in any way that affects our conclusions in this chapter. The article by Divine (2014) is a high-level discussion of intelligent design and algorithmic information theory.

For a discussion of the history of the concept of gene duplication and divergence, have a look at the article by Taylor and Raes (2004). They make it clear that the concept has been known to biology going back to the early twentieth century, making it all the more remarkable that anti-evolutionists persist in their charge that known processes cannot account for information growth in the genome.

In Section 6.3, I speculated about the possibility of defining a measure of "functional information." An attempt in that direction was proposed in a paper by Hazen, Griffin, Carothers, and Szostak (2007). Roughly, they define the functional information of a system to be the probability that a randomly chosen configuration of its parts will be able to perform the same function as well as or better than the actual configuration. This is an interesting idea, but it is unclear how useful it is for practical biological problems. In their paper, the authors apply their concept to symbolic systems such as letter sequences and artificial life experiments, as well as to certain RNA polymers.

As a mathematician, it is not exactly part of my daily routine to keep up with the literature in the field of molecular evolution. Nonetheless, I had no doubt what I would find when I started my research for Section 6.6. Molecular evolution has been a major field of study since at least the 1960s. If the only thing researchers ever discovered was that proteins are exquisitely sensitive to the slightest change, the field would have died out long ago. It was the work of about an hour, using an internet search engine and browsing through relevant academic journals, to acquire a large stack of papers showing that protein space allows for considerable exploration through known mechanisms. In reading anti-evolutionist literature, you always have to keep in mind that their descriptions of current work rarely have much connection to reality.

Throughout this book, I have accepted the premise that protein space is incredibly vast, and that only a small portion of it has been explored. I have argued that even from this starting point the anti-evolutionist arguments based on search and probability do not work. However, both of these premises have been strongly challenged. A fascinating article by Dryden, Thomson, and White (2008) provide reasons for skepticism about both premises. They write:

> Two assumptions are generally made when considering the molecular evolution of functional proteins during the history of life on Earth. Firstly, the size of protein sequence space, i.e. the number of possible amino acid sequences, is astronomically large and, secondly, that only an infinitesimally small portion has been explored during the course of life on Earth. ... As will be described below, others have concluded that the first assumption is incorrect, and we agree with this conclusion. However, we also conclude that the second assumption is incorrect and calculate that most of the sequence space may have been explored.
>
> *(Dryden, Thomson, and White 2008, 953)*

This conclusion is based in large measure on the extensive functional redundancy of proteins. That is, proteins can tolerate a lot of substitutions in their chain of amino acids. We should also take note of their closing sentence, "Hence, we hope that our calculation will also rule out any possible use of this big numbers 'game' to provide justification for postulating divine intervention." (955) They refer specifically to work by Dembski, as well as to work by ID proponent Walter Bradley, as the target of this remark.

There are many issues with the argument of Dembski's *No Free Lunch* beyond those discussed here. The article by Elsberry and Shallit (2011) is especially good. See also the lengthy review by Richard Wein (2002). I published my own review of the book in the academic journal *Evolution* (Rosenhouse 2002b). The article by Häggström (2007) also offers much food for thought, and argues that the No Free Lunch theorems are not as important to problems in combinatorial search as is sometimes claimed.

Computers have been used to simulate evolution since the late 1960s. The article by J. L. Crosby (1967) is a fascinating historical document about the early days of this research. In Chapter 4, we saw Marcel-Paul Schützenberger object to the paucity of computer simulations of evolution, but such work was already underway even then.

The paper by Meester (2009) expresses sympathy for the ID position on the narrow topic of whether the NFL theorems have implications for evolution, while stopping short of endorsing ID itself. In their reply, Blancke, Boudry, and Braeckman (2011) argued compellingly that Meester was mistaken.

For an entertaining and informative survey of the surprising successes of artificial life experiments, have a look at the article by Lehman, Clune, Misevic, et al. (2020). The book by Forbes (2004) is a very readable presentation of how biological insights have motivated developments in computer science. This includes extensive discussions of evolutionary algorithms and artificial life. For a more technical introduction to this area of research that includes extensive historical material, have a look at the book by Fogel (1999).

In my discussion of the Marks, Dembski, and Ewert book *Introduction to Evolutionary Informatics* (2017) in Section 6.10, I remarked that it was not necessary to parse the dense mathematics they use in support of their formalism since our sole interest was in their application of this formalism to biology. In a series of web posts, biologist Tom English comes to the same conclusion as me, that their work has no relevance to evolution, but also undertakes the hard work of discussing their mathematics (English 2017). See the relevant entry in the bibliography for the first of these posts, which also contains links to all of the others. The critical review by Randy Isaac (2017) also has many insightful things to say. I published my own review of the book as part of a lengthy essay about three then recent books relating mathematics to the fundamental soundness of evolution (Rosenhouse 2018). That essay contains a fair amount of material I ended up not including in this book.

7 Thermodynamics

We now turn to the most ambitious argument in the annals of mathematical anti-evolutionism. I am referring to the claim that evolutionary theory is in conflict with the second law of thermodynamics. Roughly, the claim is that evolutionary theory requires us to believe that purely natural forces have caused organisms to become more complex over time, but the second law says that this is impossible. For convenience, I shall refer to this as the "the second law argument."

In various forms, this argument has been a mainstay of anti-evolutionism since at least the 1940s. I call it ambitious because, even more so than the arguments we have discussed to this point, it makes almost no contact at all with the facts of biology. People who put forth this argument are basically saying, "We don't even have to look at your circumstantial evidence. Just stick your fossils in a museum somewhere. You can pile up genetic and anatomical comparisons all day long. The facts of embryology and biogeography may be fascinating, but they are irrelevant. Your theory contradicts the second law. End of story."

It is certainly true that the second law has an exalted status in science. Physicist Arthur Eddington famously said,

> The law that entropy always increases holds, I think, the supreme position among the laws of Nature. If someone points out to you that your pet theory of the universe is in disagreement with Maxwell's equations – then so much the worse for Maxwell's equations. If it is found to be contradicted by observation – well, these experimentalists do bungle things sometimes. But if your

> theory is found to be against the second law of thermodynamics, I
> can give you no hope; there is nothing for it but to collapse in
> deepest humiliation.   *(Eddington 1929, 74)*

In light of this status, charges of violating the second law are serious business.

However, scientists and philosophers are all but unanimous in finding the second law argument to be exceedingly poor. In their view, it is at the same level as arguing that since gravity always pulls things down, birds and airplanes are impossible. More than that, they see the argument as really so silly that its persistence in anti-evolutionist discourse just proves the utter lack of good faith on the part of their opponents. You either understand the second law or you do not, they argue, and if you do then you also understand that it does not contradict anything put forth in evolutionary theory.

I agree with this view. Still, I have a few reasons for discussing the second law argument at such length.

One is that I have seen how rhetorically powerful it can be. It was a mainstay of the creationist conferences I attended, and I often had audience members fling it at me in casual discussions after the main presentations. They did not think it was silly, to put it mildly.

Also, though thermodynamics is generally considered to be a branch of physics, the second law has a strongly mathematical character that justifies its inclusion in this book. I have not always been satisfied with the way biologists have responded to this argument, precisely because they have not taken adequate note of the underlying mathematics.

Finally, and on a more positive note, thermodynamics is fascinating, and coming to understand why the anti-evolutionist version is such a ridiculous caricature can help us appreciate the real thing. We shall have to devote quite a few pages to laying out the basics of the subject before we can turn to the arguments of the anti-evolutionists. Hopefully this material will be sufficiently interesting to make it worth the effort.

That said, even before coming to the scientific details we can say that the second law argument bears a heavy burden, since there is something implausible about it right from the start. The basic ingredients of evolution are empirical facts: genes really do mutate, sometimes leading to reproductive advantages, and natural selection can string together several such mutations into adaptive change. On a small scale, this has all been observed. But if small evolutionary changes are observable on short time scales, then it is hard to believe that an abstract principle of thermodynamics is going to rule out larger changes over longer time scales.

As we consider the many anti-evolutionist versions of the second law argument, we shall have to attend to how they attempt to circumvent this point. We shall see they have no convincing way of doing so.

7.2   WHAT IS ENTROPY?

Most people know that the second law has something to do with entropy, and that entropy has something to do with randomness and disorder. Knowing only this much, however, can be very misleading.

The problem is that it is very hard to define what entropy is. With most of the quantities you learn about in a physics class, it is easy to understand in a general sense what is meant, even if the terms can be hard to define precisely. For example, it is not so easy to define "mass," but you feel like you know what it means to say a large object is more massive than a small object. Likewise for terms like "velocity," "acceleration," or "momentum." We have enough experience with physical systems to know, at least in a track one sense, what is being discussed when these words are used.

This problem is more acute in thermodynamics since everyday terms like "heat" or "temperature" receive technical definitions that differ from their everyday meanings. Textbooks in this area devote whole sections to explaining what these terms mean, and to philosophical questions such as how to define an appropriate temperature scale. Still, you never feel completely adrift. When someone refers to heat or temperature, you feel like you know what is meant. A hot object is radiating something that a cold object is not, and that something is heat. Likewise, in normal discourse there is nothing confusing in the statement that a hot object has a higher temperature than a cold object.

Still another important thermodynamical term is "internal energy," and here again we feel we understand what is meant. If I heat a pot of water, then I am causing the water molecules to move faster, and this constitutes adding energy to the system. A system in which the molecules are moving very quickly has more internal energy than a system in which the molecules are moving slowly. Got it! In concrete situations, it might be tricky to take proper account of all the different forms of energy and their various interconversions, and this is why first-year physics students spend a lot of time working out difficult textbook problems. That notwithstanding, it is readily understood that there is a certain property of a physical system that is captured by the term "internal energy."

This brings us to "entropy." What is it? If someone brings you a thermodynamical system and says, "Show me the entropy!" you would hardly know how to reply. Unlike our other physical quantities, there is no obvious, macroscopic aspect of the system at which to point. Given this, the best way to understand entropy is to recount some of the history that went into its formulation.

Thermodynamics was born from the industrial revolution in the early nineteenth century. Heat engines of various designs had come into widespread use, leading to a wealth of practical experience in the conversion of heat into mechanical work. A common example of a heat engine in use at that time was the steam engine. Heat was added to a reservoir of water, turning it to steam. The steam would then put pressure on a piston causing it to move, which in turn led to other mechanical work, such as the revolution of a wheel. As a result of doing work the steam cools. It is then passed through a condenser of some sort, which returns the steam to a liquid state. Then the process begins anew.

It was the universal experience of engine designers that far more energy was needed to power a heat engine than was produced by the engine itself. This led to theoretical investigations into the maximum efficiency attainable by an ideal heat engine. At that time, heat was viewed as a fluid, called "caloric," and this point of view is captured in our use of the term "thermodynamics," which means "the motion of heat." The early pioneers in this area reasoned that if heat was a fluid, then a proper science of thermodynamics might begin with analogies to fluid dynamics.

Water is an especially common fluid, and water wheels were a standard technology of the time. The idea was to position a wheel at the base of a waterfall. The kinetic energy of the falling water would hit the wheel, causing it to rotate, and this rotational motion could then be converted to other sorts of mechanical work. It was quickly realized that the amount of work that could be extracted in this way was proportional to the height difference between where the water started and where it ended. The greater the height, the greater the amount of work. Moreover, the water only naturally flows one way, from the higher pool to the lower. It will not flow from the lower to the higher unless energy is expended to make it happen.

Again, it is the height difference that determines how much energy can be extracted to perform mechanical work. The water sitting in a stagnant pool at the base of the fall still contains potential energy. One could imagine excavating the land so that this water would fall through another height to a lower level still. The point, however, is that an excavation of that kind is necessary. The kinetic energy of the falling water is available to do work, while the energy of the stagnant pool is not available.

The analogous statements in thermodynamics are that a heat engine needs a temperature difference to do work, and the greater the temperature difference the greater the amount of work that can be done. Moreover, since heat naturally flows one way, from hotter to colder, the temperature gradient can only be maintained if energy is expended to maintain it. Left to nature, heat will be lost to the environment, the temperature gradient will decrease, and more and more energy will become unavailable for work.

And *that* is where entropy enters the picture: it measures the growth of the unavailability of energy to do mechanical work. More entropy means less available energy. Seen in this way, you do not so much point to some macroscopic property of a thermodynamical system and say, "There's the entropy!" Instead you typically think in terms of the change in entropy as the result of some thermodynamical process. Assuming the system is isolated from its surroundings, more and more of its internal energy will become unavailable for work. You might even say that entropy describes the transformation of energy from available to unavailable states, and, indeed, the word "entropy" comes from the Greek word for "transformation."

You will have noticed that at no point in this discussion have we said anything about order or disorder. Entropy has to do with the availability of energy to do work, and not with any everyday notions of complexity or structure. Still, you can see where someone might get that impression. The tenor of our discussion tracks well with our everyday experience that things seem to run down unless energy is expended to prevent them from doing so.

That is why when scientists try to communicate the gist of the second law to audiences who would not be receptive to a heavily mathematical treatment, they often rely on everyday examples to make their point. They will note that a room becomes dirty and disordered unless constant effort is expended to keep it clean and tidy. They will note that you cannot unscramble an egg, at least not without a considerable expenditure of energy. Or they will note that a glass dropped to the floor shatters into many pieces, but a film showing the pieces rising from the floor and reassembling into a glass is immediately recognized as something unnatural.

This is all perfectly acceptable, so long as we are content with a track one understanding of the second law and do not need to think carefully about the technical minutiae. Thinking in terms of order and disorder captures something important about the second law, despite being a crude simplification of how physicists think about it. But if someone in the audience wants to engage in serious scientific discussion, especially if they have it in their heads that thermodynamics can be used to refute a major, successful, biological theory, then a general understanding is insufficient. They will have to engage with the track two version, and that means considering some of the underlying mathematics. We turn to that in the next section.

7.3   THE FIRST TWO LAWS OF THERMODYNAMICS

Let us turn now to some track two considerations. Textbook discussions of the first two laws of thermodynamics are typically draped across many, dense, notation-filled pages. We will only need a small taste of that here. As always, it will not be necessary to parse every detail.

The earliest formulations of what we now know as the first two laws of thermodynamics arose as generalizations from experience. It was just an empirical observation that energy never seemed to be either created or destroyed, though it could certainly be changed from one form to another. It was likewise an empirical observation that heat always moved from hot to cold and never the other way around. You can build refrigerators and air conditioners to force heat to flow the other way, but a flow from cold to hot never happens spontaneously.

The notion that energy could neither be created nor destroyed came to be known as the first law of thermodynamics, and the notion that heat only spontaneously travels from hot to cold came to be known as the second law.

As the study of thermodynamics shifted from practical concerns to abstract modeling, mathematical formulations of these two laws became possible. However, before we can state these formulations, we must first make distinctions among three different kinds of thermodynamical systems.

When we speak of such a system, we have in mind some little piece of the universe that is separated from the rest by a clear boundary. Everything inside the boundary is "the system," and everything outside the boundary is "the surroundings." We then say that a system is open if both matter and energy are crossing the boundary; we say that it is closed if energy, but not matter, is crossing the boundary; and we say that it is isolated if neither matter nor energy is crossing the boundary.

A standard example is an uncovered pot of water sitting atop a lit stove. If we take the pot and the water to constitute "the system," then energy is entering the system from the flame below, and matter is leaving the system in the form of steam. This is an open system. If we put a lid on the pot then we have a closed system. Energy is still crossing the boundary from below, but matter is no longer crossing the boundary above the pot. To create an isolated system, we would need to extinguish the flame and insulate the pot thoroughly to keep it from radiating heat across the boundary. It is impossible to create an isolated system in a laboratory, and it is often said that the only truly isolated system in nature is the universe taken as a whole. However, we can certainly create a close approximation to an isolated system.

Our pot analogy is also useful for illustrating another thermodynamical concept, which we will need later in this chapter. Imagine that we have extinguished the flame and have allowed the pot to sit untouched for a lengthy period of time. The pot will radiate its heat to the surroundings. As a result, the pot cools and the surroundings warm. Eventually the two will reach the same temperature, and no further energy exchange between the pot and its surroundings occurs. At this point, the pot is said to be in "equilibrium" with its surroundings.

We can now readily produce a mathematical version of the first law. We imagine a closed system in which energy, but not matter, might be crossing the boundary. It is traditional to use U to denote the internal energy of the system, Q to denote the quantity of heat entering the system, and W to represent the work done by the system on the surroundings. We use the Greek letter Δ to denote "change in." Then we can write:

$$\Delta U = Q - W. \tag{7.1}$$

In words: the change in internal energy is found by adding the energy entering the system as heat and subtracting the energy leaving the system as work done on the surroundings.

If we restrict our attention to an isolated system, then nothing whatsoever is crossing the boundary. In this case we have $Q = W = 0$, which implies that $\Delta U = 0$ as well. Thus, in this case, the first law reduces to the statement that the internal energy of an isolated system must remain constant, which is equivalent to saying that energy can neither be created nor destroyed. In this form, the first law is often referred to as the law of conservation of energy. (It is possible to formulate versions of the first law that apply to open systems, but this requires a level of technical detail that is well beyond anything we want to discuss here.)

Whereas the mathematical formulation of the first law does not require anything especially fancy, the second law can only be expressed using some ideas from calculus. Furthermore, we will need one more bit of thermo-jargon to express it properly. It is worth the trouble to do this, so I will ask you to bear with me for a moment.

Recall that the earliest theorizing about thermodynamics occurred in the context of practical questions about heat engines. The tendency of heat to travel from hot to cold, and not vice versa, was the main principle underlying the functionality of such engines, and we saw that engines tend to dissipate energy into less available forms. This led to practical questions about how efficient a heat engine could possibly be.

It was quickly realized that, among other attributes, the most efficient possible heat engine would be one in which no energy was lost to the surroundings, but was instead converted perfectly into work. Real heat engines, no matter how well designed, inevitably waste some energy because of friction between moving parts or by radiating heat to the surroundings, but an ideal heat engine would not have these defects.

This led to the notion of a "reversible" process. The idea is best explained by example. Imagine a piston in a cylinder that compresses a small quantity of gas. If we add a small quantity of heat to the system, then the gas expands and pushes the piston outward. But if we now cool the gas to precisely its original temperature, the piston returns precisely to its original position. This is an example of a reversible process. A small change to the surroundings leads to a small change to the system, and restoring the surroundings to their original state restores the system as well.

Real thermodynamical processes are never perfectly reversible precisely because some energy is always lost to the surroundings. In a real process, restoring the surroundings to their initial state does not perfectly restore the system.

Therefore, the concept of "reversibility" captures the idea that an ideal heat engine converts all of its heat to work, and it is for this reason that entropy calculations are always carried out under the assumption that we are studying a reversible process. More precisely, in calculating the change in entropy that occurs when a system goes from some initial state to some final state, we begin by describing a reversible process taking us from one state to another, and then we evaluate a certain mathematical expression.

Here is that expression, where S represents entropy, Q represents the heat added to the system, and T represents the temperature of the system when the heat is added. (For technical reasons, it is important to use a temperature scale on which $0°$ represents absolute zero, but this detail is not important for our purposes.)

$$\Delta S \geq \int_{\text{rev}} \frac{dQ}{T}. \tag{7.2}$$

The expression dQ represents a small input of heat into the system, and the integral sign (the elongated S) indicates that we find the total entropy change by adding up the numerous small entropy changes that arise in various parts of the system as we go from the first to the second state. The "$\geq$" sign means "greater than or equal to." It turns out that the two sides are equal only in the context of a purely reversible process. Since perfect reversibility is an ideal to which any real heat engine can only aspire, we conclude that in any actual process the change in entropy will be greater than what is given by the right hand side.

Equation 7.2 applies to any system, whether open, closed, or isolated. However, as with our discussion of the first law, the equation becomes easier to understand if we restrict our attention to an isolated system. In that case, neither matter nor energy are crossing the boundary. Therefore, the integral evaluates to 0, and the second law reduces to the statement that the change in entropy will be greater than 0, which is to say there will be more entropy at the end of the process than there was at the beginning. Hence the assertion that in an isolated system, entropy cannot spontaneously decrease.

If the system is open, then there is the possibility that the integral on the right hand side evaluates to something negative. To say that the change in entropy is negative is to say that there is less entropy in the final state than there was in the initial state, which is to say the entropy has decreased. Thus, the second law allows for spontaneous entropy decreases in open systems, but even in such cases it precisely quantifies just how much of a decrease is possible.

As we have said, in a typical thermodynamics textbook, equation 7.2 appears after a lengthy and difficult mathematical derivation. That level of technical detail is well beyond anything we will need for the discussions to come. However, it is worth taking a moment to understand, in general terms, why dividing heat by temperature is a plausible way of measuring entropy.

Remember that with fractions, the bigger the top the bigger the fraction. On the bottom it is the reverse: the smaller the bottom the bigger the fraction.

Thus, for a fixed value of T, our formula says that a large influx of heat will increase the entropy more than a small influx of heat. If we think of entropy as a tendency for energy to become unavailable for work, then we are basically saying that an engine operating at high heat will burn itself out faster than an engine operating at low heat. Adding heat to the engine will accelerate the pace at which energy becomes unavailable for work, and that is why warming a system increases its entropy.

On the other hand, what if we imagine keeping the top of the fraction constant and only varying the bottom? In that case, our formula says that a given quantity of heat will cause greater entropy increase if the system is at a low temperature than if it is at a high temperature. To see why this makes sense, it is simplest to think of entropy as a tendency toward disorder and then to argue by analogy. If you are in a silent library and you suddenly clap your hands, you will create a serious disturbance. The other library patrons will be very annoyed with you. But if you clap your hands on a busy street corner, probably no one will notice the small amount of extra noise. Alternatively, imagine that you light up a 40 Watt bulb in a dark room. The result is a dramatic increase in visibility. If you illuminate the same bulb in a room that is already brightly lit, then you will barely notice the extra light.

Likewise, a small amount of heat added to a system at a low temperature is more disruptive than the same amount of heat added to a system at a high temperature.

Now, the discussion of the last few paragraphs has been a bit technical. I promise I will not be using equation 7.2 to actually calculate anything, which is something I have in common with the anti-evolutionists whose work I will be criticizing. However, I felt it was important to spend some time dwelling on real thermodynamics, just so that we could better understand why the anti-evolutionist version is so hard to take seriously.

The *really* important part of our discussion is not so much the mathematical expression itself, but rather the precision of the result.

The second law of thermodynamics is a precise mathematical statement that exactly quantifies what is possible and what is not. When this is understood, you realize why in serious scientific discourse, as opposed to casual conversation, it does not make sense to say that a proposed physical process *appears* to violate the second law. Either it does or it does not, and if you claim it does then it is on you to provide the calculation that shows that it does.

In other words, show me that the process entails an entropy change that is smaller than what is required by equation 7.2. If you are not prepared to do so, then you are not making a serious argument and should not be talking about thermodynamics at all.

7.4   STATISTICAL MECHANICS

Based on what I have said up to this point, you would never know that matter is made up of microscopic atoms. Instead, I have based my presentation on the macroscopic properties of matter, which we can define roughly as those properties that are visible and measurable to the naked eye. Since this presentation tracks well with how matter was classically viewed prior to the ascension of atomic theory in the nineteenth century, it is typically referred to as "classical thermodynamics."

If instead we take the view that any given bit of matter is a large assemblage of microscopic particles, we can gain some further insight into our thermodynamical principles. This leads to a branch of physics known as "statistical mechanics." It involves "mechanics" in the sense that we are interested in the motions of particles, and it is "statistical" in the sense that we are not interested in the motion of any specific particle, but only in the average behavior of large ensembles of particles.

To illustrate the main idea, it is customary to imagine a sealed box with a partition in the middle. We imagine that there is a gas on one side of the partition and vacuum on the other side. When we remove the partition, the gas molecules rush to fill the vacuum, and before long they are distributed evenly throughout the box.

It never happens that the evenly distributed gas molecules suddenly rush to one side of the box, leaving a vacuum on the other side. This is reminiscent of the macroscopic observation that heat only travels in one direction.

The behavior of the gas molecules is readily understood in terms of probability. We can assume there are so many molecules moving and colliding in effectively random ways that any particular arrangement of the molecules is as likely as any other. Since there are vastly more arrangements in which the molecules are evenly distributed throughout the box than there are arrangements where all the molecules are found on one side, we are far more likely than not to find the molecules evenly distributed.

Viewed in this way, we can say that the standard variables of classical thermodynamics arise as the result of the average behavior of large ensembles of particles. We can define the "microstate" of the gas to be the position and momentum of each individual molecule. We can contrast this with its "macrostate," which is characterized by familiar variables like internal energy, pressure, volume, and temperature. The microstates in which the gas molecules are evenly distributed all give rise to the same macrostate. Conversely, macrostates in which most of the energy is unavailable for work are consistent with many microstates, and that is why they are more likely to occur naturally than are macrostates with a lot of available energy, which are consistent with only a small number of microstates.

We can also formulate a statistical notion of entropy, and the standard formula for doing this entails that increasing the number of microstates increases the entropy. Heating the gas increases the velocity of all the particles, making possible more vigorous collisions than before, and this increases the number of microstates available to the molecules. In this way, we can understand why heating a system increases its entropy. The second law can be understood as the statement that since microstates corresponding to disordered states of matter vastly outnumber microstates corresponding to ordered states, disordered states are far more likely to occur naturally than ordered states.

The statistical view provides insight into thermodynamics that is not available under the classical view. However, something has been lost as well. Viewed statistically, the second law is now a probabilistic statement about what is exceedingly likely to happen, as opposed to a precise mathematical statement of what will definitely happen.

The statistical view is more modern, and it is more fundamental in the sense that we are explaining the behavior of macroscopic systems by referring to the collective behavior of the particles that compose them. It should not be thought, however, that the statistical view has somehow supplanted the classical view or has shown that the classical view is obsolete. Rather, the classical and statistical views are just two different approaches to the same problems, each useful in its own domain. H. C. Van Ness, a chemical engineer, provides an apt summary of their relationship:

> Statistical mechanics adds to thermodynamics on its theoretical side, as a means for or as an aid to the calculation of properties. The other half of thermodynamics, the applied half, benefits only from a wider availability of the data needed in the solution of engineering problems. ... [I]t does provide, as thermodynamics does not, the means by which thermodynamic properties may be calculated whenever detailed descriptions of atomic and molecular behavior are provided from other studies, either theoretical or experimental. Thus statistical mechanics adds something very useful to thermodynamics, but it neither explains thermodynamics nor replaces it. *(Van Ness 1969, 103)*

When presenting the second law argument, anti-evolutionists sometimes use classical thermodynamics and sometimes use statistical mechanics. More often, they are vague in their presentation and leave it to the reader to guess which is intended. At any rate, I have belabored the statistical version here since we shall see it again later in the chapter.

7.5   APPLYING THE SECOND LAW TO EVOLUTION

At this point you might be wondering: What does any of this have to do with evolution? The succinct answer is that it has very little directly to do with it.

There are two key principles arising from our discussion. The first is that thermodynamics is about bringing mathematical precision to questions about heat flow. The second is that likening entropy to disorder is sometimes useful when trying to communicate the flavor of the second law to lay audiences, but it is not correct when applying the second law to serious scientific questions.

If the anti-evolutionist's argument does not employ the mathematical machinery of thermodynamics, if instead it is based on nothing more than the vague observation that things naturally decay and break down while evolutionary theory says they have become more complex over time, then there is no reason to invoke the second law at all. Referring to the second law can be rhetorically useful, since it creates an aura of scientific precision, but it adds nothing to the substance of the argument. One can say simply that an increase in ordered complexity throughout natural history is not the sort of thing experience tells us to expect. Any biologist suggesting such a thing had better have a good explanation for how the growth in complexity occurred.

Of course, biologists reply that they do have such an explanation. Passing chance genetic variations through the sieve of natural selection, can, over time, lead to great increases in complexity, as we have discussed. Anti-evolutionists reject that explanation, but the point is that the ensuing debate has nothing to do with thermodynamics.

However, we should give some attention to the second principle. Entropy change has far more to do with heat flow into and out of a system than it does with everyday notions of order and complexity.

For example, imagine once more the pot of water on a stove. The flame radiates heat that is absorbed by the pot. As a result, the flame cools slightly and its entropy decreases. The pot is heated, and its entropy increases. The second law implies that the entropy increase of the pot must be greater than the entropy decrease of the flame.

Let us apply this to the sun and earth. The sun radiates energy to the earth, and consequently its entropy decreases, if only slightly. The earth is warmed and its entropy increases. Nearly all of the heat sent by the sun to the earth is then radiated back into space. This tends to lower the entropy of the earth and to increase the entropy of space. The radiation from the earth into space takes place at a much lower temperature than the radiation from the sun to the earth, and this implies that the entropy increase due to the former radiation more than outweighs the entropy decrease due to the latter. In other words, if we take the earth, sun, and some quantity of the surrounding space as our system, then we can say that the entropy decrease of the sun is more than outweighed by the entropy increase of the surrounding space.

Presumably, there has also been some entropy decrease on earth as a result of the evolutionary process. If we are thinking casually about entropy as a statement about disorder, then it seems plausible that functioning animals represent lower entropy states than the same molecules lying in a lifeless pile. It would be nice to quantify the entropy decrease due to evolution, but there are grave difficulties that make it all but impossible to do so. Since this point will arise naturally in Section 7.8, we shall defer further discussion until then.

Physicists often present the second law as the statement, "In an isolated system, entropy cannot spontaneously decrease." As we have seen, this statement follows as a special case of equation 7.2. Formulated in this way, the second law makes no assertion whatsoever about entropy change in open systems. Biologists are quick to point this out, noting that the earth is an open system since it is bathed in energy from the sun, and therefore simplistic applications of the second law to evolutionary theory are erroneous. Here is a representative quotation, from Richard Dawkins:

> When creationists say, as they frequently do, that the theory of evolution contradicts the Second Law of Thermodynamics, they are telling us no more than that they don't understand the Second Law (we already knew that they don't understand evolution). There is no contradiction, because of the sun!   *(Dawkins 2009, 415)*

Assertions of this general sort are commonplace in the anti-creationist literature.

This is a perfectly acceptable reply, and it certainly exposes a clear error in simplistic versions of the second law argument. However, we have seen that the restriction to isolated systems is more properly viewed as a special case of the second law. If the second law is construed as the relationship expressed in equation 7.2, then it applies to any sort of system. In light of this, let us finally turn the floor over to the anti-evolutionists, to better understand why they think the second law poses a challenge to evolutionary theory.

7.6   THE BASIC ARGUMENT FROM THERMODYNAMICS

We shall take for our starting point a pair of papers from the early 1940s. One was published by R. E. D. Clark in 1943, while the other was published by E. H. Betts in 1944 (Clark 1943, Betts 1944). Both were presented to the Victoria Institute, an organization of Christian scientists created shortly after the publication of Darwin's *On the Origin of Species*, and which, under the working name "Faith and Thought," remains in existence today. Interestingly, both papers were called "Evolution and Entropy." Since both authors presented similar arguments, while Clark's were presented with greater cogency and lucidity, we shall use his paper as representative.

Throughout his essay, Clark refers to the "law of entropy," instead of the "second law of thermodynamics." Though this terminology is seldom used today, for the purposes of this section we shall mostly follow Clark's usage.

Clark is clear that his argument is meant as an appeal to common sense, but one that is backed up by notions from physics. He notes with some asperity the willingness of philosophers to challenge our most familiar and fundamental metaphysical constructs, and replies thus:

> Every intuitive idea has suffered a similar fate; philosophers have doubted causality, have doubted the existence of a physical world, have doubted interaction between mind and matter, have doubted every conceivable dictate of common sense. And so, right up to modern times, men and women are to be found who suppose that by the working of some inscrutable principle, nature is in the habit of producing order where chaos existed before. *(Clark 1943, 50)*

He is equally clear that in speaking of the "law of entropy" he envisions something more general than what physicists and engineers typically have in mind:

> It seems advisable, therefore, to extend the meaning of the word "entropy" so as to make it a synonym for "disorder." In this sense the "law of entropy" must be understood to mean the law that disorder will tend to increase, but that order can never arise spontaneously from chaos. It is in this form only, of course, that the law is related to the theory of evolution. *(Clark 1943, 51)*

In a footnote to this passage, Clark explicitly contrasts his version of the law with, "the specialised law of entropy of the physicist and engineer ..." (Clark 1943, 51)

He continues the argument by suggesting that in light of this proposed law, apparent increases in order must be explained away as being illusory:

> In the world of physics chaos is constantly increasing, energy is becoming less and less available. But while some of the still ordered energy is turning into energy in a less ordered condition, it will chance now and again that groups of atoms will arrange

themselves in what *appear* to be new ways. It will seem to the
uninitiated as if atoms and molecules have arranged themselves
and created something new; but the scientist tries to show that
however startling the novelties that emerge, they were really
present all the time: they are the logical and deducible
consequences of what was already in existence.

(Clark 1943, 52, italics in original)

These passages illustrate how Clark understands the the law of
entropy. The connection to evolution now seems clear. The dramatic
appearance of complex organ systems in biology in the course of natural history does not seem like the unfolding through natural law of an
already existing order. Rather, such appearances constitute genuine
novelty and genuine increases in complexity. Clark's conclusion:

It seems reasonable to conclude, therefore, that if in past ages
complex organisms ever did evolve from simpler ones, the process
took place contrary to the laws of nature, and must have involved
what may rightly be termed the miraculous.   *(Clark 1943, 63)*

We have now seen enough of Clark's argument to offer some
initial replies.

We can think of Clark's offering as the basic argument from
thermodynamics: Evolutionary theory says that animals have naturally gotten more complex over time, but the second law says things
naturally break down. This is a contradiction.

In this form, we have a distinctly track one argument. Clark
is admirably forthright that his is an argument based on "common
sense," and he is explicit that his use of the law of entropy differs
from how physicists would use it. Therefore, the basic argument is
highly vulnerable to the point we made at the end of Section 7.3. Since
we are not using any of the mathematical machinery underlying the
second law, thermodynamics is only playing a rhetorical role in the
argument. As we have noted, you do not need to invoke anything from

physics to justify the claim that things tend to break down unless energy is expended to maintain them.

We could stop there and just move on to the next section, but there are a few other aspects of Clark's argument that are worth commenting on. He apparently believes that our everyday understanding of "order" and "disorder" is sufficient for grasping his intent. He goes on to urge that the law of entropy, though used by physicists and engineers to apply only to certain sorts of physical systems, be extended to become a general principle of increasing disorder.

However, there is a reason why physicists themselves have not made this move. As we have seen, the law of entropy as understood in scientific discourse is a mathematical statement. The terms "order" and "disorder" never appear in such discussions, since there is no way to define them with sufficient precision. Everyday notions of order and disorder are frequently in conflict with the technical understanding of entropy.

A simple example is to compare a bowl of liquid water to the same volume of water in the form of jagged ice shards. Intuitively, we might think that the smooth, homogeneous water is more ordered than the jagged ice. However, the ice is at a lower temperature and therefore has the lower entropy. Moreover, our usual notions of order and disorder involve arrangements of physical parts, while entropy is about energy flow and dissipation. Examples such as these show why it is an error to treat entropy and disorder as synonymous.

The basic argument also seems to be vulnerable on another front. It is not difficult to find examples where order, in the everyday sense, spontaneously increases, such as in the formation of a snowflake. Clark dismisses all such examples as instances of latent order being expressed, as opposed to the creation of something genuinely novel. But this distinction – between latent order and genuine novelty – is hopelessly vague. How do we distinguish "latent order" from "genuine novelty"? Clark discusses crystallization, and makes much of the fact that the crystal's structure is entirely determined by the physical properties of the system in which it emerges. In his telling, it is specifically this physical determinism that places crystallization in the category of latent order. However, this understanding of the distinction is clearly insufficient for his argument because small increases in order can easily arise just by chance. Therefore, we must also allow for events that occur as combinations of chance and physical laws.

At this point we have to wonder which part of the evolutionary process Clark believes to be impossible. After all, we have already noted that on a small scale, evolutionary innovation is seen to occur. If Clark would dismiss such innovation as an expression of latent order, then the entire panoply of biological evolution can be similarly dismissed. In other words, using Clark's understanding of the terms, we might suggest that no genuine novelty has actually arisen in the course of evolution. It has all been a mere expression of order latent in the genomes of the earliest life forms and the environments in which they found themselves.

That this seems like an unhelpful way of looking at things suggests a weakness in Clark's proposed law, at least to the extent that he seeks to apply it to evolutionary theory.

Clark briefly addresses the possibility of evolution by natural selection:

> Julian Huxley attempts to avoid the difficulty by invoking natural selection. "Natural selection," he writes, "achieves its results by giving probability to combinations which would otherwise be in the highest degree improbable. This important principle clearly removes all force from the 'argument from improbability' used by many anti-Darwinians, such as Bergson." But molecular combinations are not made more probable if, when once they have been formed, they are enshrined in a species. The analogy of the crystal nucleus shows us the extreme limits of spontaneous ordering in nature, and it is an analogy which is unfavourable to the mechanistic evolutionist.
>
> *(Clark 1943, 60)*

Some historical context might help to understand Clark's point, since we must recall that he was writing in 1943. By the start of the twentieth century, the scientific community was united in its acceptance of common descent, but divided over the mechanism of evolution. A school of thought known as "mutationism" held that mutations alone were the fundamental mechanism. Proponents of this view did not accept Darwin's emphasis on evolution's gradualness. They argued instead that the process unfolded in large, discrete steps, via what we would today refer to as "macromutations."

By 1943 this school had largely yielded to the view expressed by Huxley in the quotation above. Clark nonetheless devotes considerable space to refuting mutationism as unworkable. He was right to be skeptical (though we should note that the problems with mutationism have nothing to do with entropy or thermodynamics). However, in emphasizing the implausibility of large-scale mutations leading to significant increases in order, Clark seems to have missed Huxley's point entirely. Clark writes as though natural selection is just an add-on to mutationism, so that selection merely preserves large-scale changes after they occur by chance macromutations. This is not the case, of course. Huxley's point was that the improbability of complex systems can be broken down into a large number of small, manageable steps, each of which is preserved by natural selection. Clark never addresses this possibility, and therefore his argument simply fails.

7.7 STATISTICAL MECHANICS VERSUS EVOLUTION?

Not long after after Clark presented his case to the Victorian Society, the French biophysicist Pierre Lecomte du Noüy published his own version of the second law argument. It appeared in his 1947 book *Human Destiny*, which was a bestseller at the time.

At its core, du Noüy's argument was the same as Clark's: evolutionary theory implies an increase in ordered complexity over time, while the second law implies this is not possible. However, he made two contributions to the argument's development. The first was to link it specifically to the probabilistic understanding of entropy provided by statistical mechanics:

> One of the greatest successes of modern science was to link the fundamental Carnot–Clausius law (also called the second law of thermodynamics), keystone of our actual interpretation of the inorganic world, with the calculus of probabilities. Indeed, the great physicist Boltzmann proved that the inorganic, irreversible evolution imposed by this law corresponded to an evolution toward more "probable" states, characterized by an ever-increasing symmetry, a leveling of energy. ...
>
> Now, we men, at the surface of the earth, are witnesses to another kind of evolution: that of living things. We have already seen that the laws of chance, in their actual state, cannot account for the birth of life. But now we find that they *forbid* any evolution other than that which leads to less and less dissymmetrical states, while the history of the evolution of life reveals a systematic *increase* in dissymmetries, both structural and functional.
>
> (du Noüy 1947, 41–42)

Later, du Noüy is more blunt:

> [T]he evolution of living beings, as a whole, is in absolute contradiction to the science of inert matter. It is in disagreement with the second principle of thermodynamics, the keystone of our science, based on the laws of chance.   (du Noüy 1947, 224)

Du Noüy's second contribution was to provide numerical precision, in the form of a calculation regarding the chance formation of a single protein molecule:

> It is impossible, because of the tremendous complexity of the question, to lay down the basis for a calculation which would enable one to establish the probability of the spontaneous appearance of life on earth. However, the problem can be greatly

> simplified and we can try to calculate the probability of the
> appearance, by chance alone, of certain essential elements of life,
> certain large molecules, proteins for instance.   *(du Noüy 1947, 33)*

If you have read Chapter 5, then you know what is coming next. The calculation proceeds along the lines described in Section 5.5, and fails for the reasons enumerated there. However, there is a further aspect of this to which we should call attention.

It is interesting that du Noüy saw his calculation as a contribution to the second law argument. That is, he did not merely say, "I have calculated a small probability, and this shows evolution is impossible." Instead he said something closer to, "I have calculated a small probability. And since the second law of thermodynamics says that low probability configurations do not occur naturally, this augurs against evolution, or at least against the spontaneous appearance of life."

Physicist Richard Burling published an eloquent rebuttal to du Noüy's calculation in 1953 (note that du Noüy based his calculation on prior work due to the Swiss physicist Charles-Eugène Guye):

> [du Noüy's] computation is spurious since it ignores relevant
> chemical knowledge and is based on the obviously simple
> assumption that atoms exert on each other forces analogous to
> those exerted by billiard balls. By ignoring any knowledge about
> the chemical forces acting between atoms, as Guye did, and
> making random arrangements of billiard-ball models of sodium
> and chlorine atoms, one can easily show in the same way that the
> exact alternation of Na and Cl atoms required for the formation of
> even a submicroscopic crystal of common salt is a statistically
> "impossible" miracle, completely unintelligible in a universe
> subject to the second law of thermodynamics.   *(Burling 1953, 184)*

Of more relevance to our present concerns, however, are the problems inherent in the second part of du Noüy's argument. Even if we were to accept for argument's sake the legitimacy of his computation, it would still not be correct to see it as supportive of the second law argument.

The structure of du Noüy's argument is clear: The chance formation of even a single protein is highly improbable, and since the second law is fundamentally about probability we have established a conflict between evolutionary theory and the second law. However, the statistical mechanical interpretation of entropy, on which du Noüy bases his argument, is not about probability per se. Rather, it is about probability theory deployed in a particular manner with regard to systems that are seen to satisfy certain properties. The founders of statistical mechanics meant to apply their reasoning to gases, where you have large numbers of identical particles that are randomly colliding with each other in the manner of billiard balls. Their methods do not apply in any straightforward way to problems in organic chemistry or to the biology of living organisms.

In short, it is the probability calculation itself that is doing all the work in du Noüy's discourse and not anything learned from thermodynamics. As with the other, similar, calculations we have previously considered, it is based on a model too unrealistic to be useful. It also provides no support to claims of a contradiction between evolutionary theory and the second law.

7.8   THERMODYNAMICS IN "THE GENESIS FLOOD"

After du Noüy, the second law argument was advocated by Henry Morris. In a series of writings starting with his 1961 book *The Genesis Flood*, coauthored with John Whitcomb Jr., Morris gave the second law a prominent place in his writing. He developed the argument in several ways beyond what we have seen.

Morris, a hydraulic engineer who spent 11 years as a faculty member at Virginia Polytechnic Institute, is generally credited with founding the modern young-Earth creationist movement. *The Genesis Flood* can be viewed as the founding document of this movement. Morris would later go on to create the Institute for Creation Research, which still exists today.

Referring to work by the Princeton physiologist Harold Blum, the substance of which we shall consider momentarily, Whitcomb and Morris write:

> Blum, impressed with the universality of the entropy principle in nature and yet believing that the world and all living things have developed by means of the supposed universal principle of evolution, has attempted in a profound and influential work to harmonize and even essentially to equate entropy and evolution. But this is an impossible task, because really the one is itself the negation of the other. Creation (or what biologists imply by "evolution") actually has been accomplished by means of creative processes, which are now replaced by the deteriorative processes implicit in the second law. The latter are probably a part of the "curse" placed upon the earth as a result of the entrance of sin (Genesis 3:17), the "bondage of decay" to which it has been "subjected" by God for the present age (Romans 8:20–22).
>
> *(Whitcomb and Morris 1961, 224–225)*

It is beyond the scope of this chapter to critique Whitcomb and Morris's theology, though this statement is illustrative of their explicitly religious style of argumentation. The emphasis on Blum's work is significant, however, since it leads us to revisit a point I raised in Section 7.5: there are grave difficulties surrounding any attempt to calculate the entropy change resulting from evolution.

In 1951, Blum published a book called *Time's Arrow and Evolution*. It was an important work, in that it was a detailed attempt to explore the connections between evolution and thermodynamics. Near the end of the book, Blum discusses the general question of whether the evolutionary process as a whole can be said to contravene the second law. In his discussion, he emphasized, as we have, that a proper calculation is needed to establish that the second law has been violated, but that such a calculation is very difficult, if not impossible, in the case of evolution. In the following quotation, Blum imagines a reader challenging him on the grounds that he has not actually shown that evolutionary theory is consistent with the second law. Blum replies:

> True enough. But the important thing is the converse of this. That is, in order to deny the applicability of the second law these magnitudes would have to be measured, and until this is done the failure of the law cannot be proven. As we pointed out earlier in the book, the principal reason for accepting the second law of thermodynamics is that it has always worked wherever it has been possible to make the necessary measurements to test it; we assume therefore that it holds where we are unable to make such measurements. *(Blum 1951, 202)*

Whitcomb and Morris are unimpressed with Blum's assertions:

> [B]lum, more than most other modern evolutionary biologists, has faced seriously the implications of the entropy principle in biological evolution. Most evolutionists have simply ignored the problem or have blandly asserted that the second law is refuted by the fact of evolution. But, as Blum insists, the second law of thermodynamics has always proved valid wherever it could be tested. He bravely proceeds, therefore, to attempt to reconcile it with that with which it is utterly irreconcilable, the assumption of universal developmental evolution! Needless to say, he fails utterly. *(Whitcomb and Morris 1961, 225–226)*

Whitcomb and Morris's bravado can easily distract us from the fact that they do not actually respond to Blum's point. Specifically, they pay no attention to the need for carrying out a calculation when making claims about the second law.

Following Blum, we have emphasized that the second law is a mathematical statement. The problem is that living systems, and the processes through which they form, are not at all the sorts of systems to which classical thermodynamics applies. We have previously discussed the distinctions between reversible and irreversible processes, as well as the distinction between systems that are near to or far from equilibrium. Whereas the classical understanding of the second law applies to reversible processes undertaken on systems that are close to equilibrium, living organisms are far from equilibrium and are formed through irreversible processes. For this reason, it is effectively impossible to carry out the relevant calculation.

Put succinctly, no one knows how to answer a question such as, "What is the change in entropy associated with evolving an elephant from simpler organisms over the course of millions of years?" To be clear, we are not saying that living creatures somehow violate the second law. Rather, the claim is that it is unclear how the concepts underlying the second law apply to the evolutionary process.

Physicist Percy Bridgman, a Nobel laureate, makes this point bluntly:

> Many of the set-ups proposed for exhibiting the relation of living things to the second law do not properly reproduce the conditions necessary for the application of the law. For instance, the environment of most living things is a stream of radiation from the sun to the earth from which they extract energy which is used in the "organization" of the environment. The stream itself is a factor with "order" in the determining conditions; to prove that the second law has been violated would demand a quantitative proof that the "order" created by the organism in the final product is greater than the order in the stream of energy which made the process possible.   (Bridgman 1941, 209)

Later he writes:

> If we could assign a definite entropy to an organism we could at once answer our question about the second law. To assign an entropy to an object demands some reversible method of getting to the object from a standard starting point, and this, for an organism, is close to the problem of the artificial creation of life.
> (Bridgman 1941, 213)

Physicist Leon Brillouin was even more blunt, writing in 1949:

> How can we compute or even evaluate the entropy of a living
> being? In order to compute the entropy of a system, it is necessary
> to be able to create or destroy it in a reversible way. We can think
> of no reversible process by which a living organism can be created
> or killed: both birth and death are irreversible processes. ... The
> entropy content of a living organism is a completely meaningless
> notion.   (Brillouin 1949, 564)

Nothing has happened in the years since Bridgman and Brillouin's work to alter their conclusions.

We have noted that, as a practical matter, it is not possible to calculate the entropy change resulting from evolution. What we *can* do is work out the entropy change associated with the radiation of heat from the sun to the earth, and then the entropy change when most of that heat is then radiated from the earth back into space.

The quantity of heat radiated from the sun to the earth can be well-estimated, and nearly all of that heat is then radiated back into space. However, the sun is at a much higher temperature than the earth. Recalling our discussions in Sections 7.3 and 7.5, the lower the temperature the higher the entropy change. That means the entropy increase associated with the earth's radiation is much higher than the entropy decrease associated with the sun's radiation. This difference would essentially put an upper bound on the allowable entropy decrease due to evolution. That is, it will allow us to say that as long as the entropy decrease due to evolution is smaller than the entropy increase associated with the radiations from the sun to the earth, and then from the earth into space, then evolutionary theory is not in conflict with the second law.

Physicist Robert Oerter filled in the numerical details of this calculation. As a point of reference, he also worked out the entropy change required to freeze the world's oceans. His conclusion:

> Now, the mass of all the living organisms on earth, known as the
> biomass, is considerably less than the mass of the oceans (by a

very generous estimate, about 10^{16} kilograms. If we perform a similar calculation using the earth's biomass, instead of the mass of the oceans, we find that the second law of thermodynamics will only be violated if the entire biomass is somehow converted from a highly disorganized state (say a gas at 10,000 K) to a highly organized state (say, absolute zero) in about a month or less.

Evolutionary processes take place over millions of years; clearly they cannot cause a violation of the second law.   (Oerter 2006)

These estimates firmly place the ball in the court of the anti-evolutionists. It is on them to provide the calculation that establishes the correctness of their claims. It seems unlikely they will ever be successful in doing so.

7.9   HENRY MORRIS'S LATER WRITING

The Genesis Flood contains only a few pages addressing the second law of thermodynamics. In those pages, Morris and Whitcomb write as though it is simply obvious that the "upward" trend of the evolutionary process is at odds with the "downward" trend required by the second law. To the extent that they made any argument at all, it was directed entirely at Harold Blum in the manner explored in Section 7.8.

In the three decades following the publication of The Genesis Flood, young-Earth creationism became the dominant form of anti-evolutionism in the United States. Henry Morris was the most prominent public face of the movement, and he produced a large number of books and pamphlets defending his point of view. Thermodynamics was a frequent topic of discussion in this work.

To be blunt, most of this writing was of such low quality that it merits consideration here only for its historical significance in the development of the second law argument. Here is a representative quotation, from Morris's 1975 book The Troubled Waters of Evolution:

> Not only is there no evidence that evolution ever *has* taken place, but there is also firm evidence that evolution never *could* take place. *The Law of Increasing Entropy* is an impenetrable barrier which no evolutionary mechanism yet suggested has ever been able to overcome. Evolution and entropy are opposing and mutually exclusive concepts. If the entropy principle is really a universal law, then evolution must be impossible.
>
> *(Morris 1982, 111)*

It was in response to this sort of childish rhetoric, then ubiquitous in the literature of anti-evolutionism, that scientists, with some impatience, noted the importance of distinguishing open from closed systems. Morris was aware of this distinction, and replied thus:

> Obviously growth cannot occur in a closed system; the Second Law is in fact *defined* in terms of a closed system. However, this criterion is really redundant, because in the real world closed systems do not even exist! It is obvious that the Laws of Thermodynamics apply to open systems as well, since they have only been tested and proved on open systems!   *(Morris 1982, 124)*

As we have seen, the expression of the second law in terms of closed systems is just a special case of a more general inequality that applies to any sort of system. To repeat, if your claim is that entropy cannot spontaneously decrease, then you had better be talking about an isolated system. If the system is open there is still a strong statement to be made about the permissible entropy change, but that statement allows for entropy increases.

Were this Morris's only contribution we could simply move on. However, he did introduce one novel claim into the discussion that is worth a look. Continuing from the last quotation, Morris claims to have discovered the precise criteria that must be satisfied for a local entropy decrease to occur. Specifically, there are four criteria. The first two are that the system be open and that there be available energy. The third is a "coded plan":

There must always, without known exception, exist a pre-planned program, or pattern, or template, or code, if growth is to take place. Disorder will never randomly become order. Something must sift and sort and direct the environmental energy before it can "know" how to organize the unorganized components.

(Morris 1982, 124–125)

Finally, he says we need an "energy conversion mechanism":

It is naively simplistic merely to say: "The sun's energy sustains the evolutionary process." The question is: "*How* does the sun's energy sustain the evolutionary process?" This type of reasoning is inexcusable for scientists, because it confuses the First Law of Thermodynamics with the Second Law. There is no doubt that there is a large enough *quantity* of energy (First Law) to support evolution, but there is nothing in the simple heat energy of the sun of sufficiently high *quality* (Second Law) to produce the infinitely-ordered products of the age-long process of evolutionary growth.   *(Morris 1982, 127)*

As is typical with anti-evolutionist writing, we must hack our way through thickets of nonsense before we come to the actual argument. When someone claims that evolutionary theory contradicts the second law, it is perfectly reasonable to reply that the earth is bathed in energy from the sun. When energy enters a system it is possible for entropy to spontaneously decrease, and such a decrease is entirely consistent with the second law. It is an interesting question to ask precisely how the sun's energy drives evolution, but that question is different from the one we started with, which was whether evolutionary theory and the second law conflict. Moreover, the first law has nothing to do with whether there is enough energy available to drive some process, but is instead about the change in internal energy of a system as the result of energy flow through it. Any confusion about the roles played by the laws of thermodynamics is entirely on the anti-evolutionist side.

The point is that Morris is conflating two separate questions. One question is, "Does evolutionary theory contradict the second law of thermodynamics?" The answer to that question is flatly no, for reasons we have already discussed. An entirely separate question is, "How does the sun's energy drive the evolutionary process?" *That* question is answered in detail in the world's biology textbooks, as opposed to the world's thermodynamics textbooks. Any question about energy flow is ultimately related to thermodynamics, but the fact remains that explaining photosynthesis and metabolism are problems for biologists and not physicists. This topic will arise again in Section 7.10, so we will defer further discussion until then.

The extreme silliness of Morris' argument can be seen from considering a parallel case. Suppose someone, let us call him John, argues that flight conflicts with the law of gravity because gravity always pulls things down. In reply, a patient physicist will point out that the law of gravity only says that any two objects with mass apply a force on each other. She will go on to explain that a sufficiently strong upward force can counteract the downward force we experience from the earth, and that is why flight is possible. But now John plays his trump card. He says, "Your reasoning is inexcusable! Sure, maybe in principle you could generate an upward force, but the question is *how* you generate such a force!"

The physicist would reply precisely as we replied to Morris. She would say, "But now you are just changing the subject. Your first point was that flight conflicts with gravity, and I explained why that is wrong. An entirely different question is the mechanism we use to make flight possible. That is an interesting and important question, but it is separate from any claim that flight and gravity conflict."

The second law argument languished in this sorry state until its modern revival at the hands of ID proponents.

7.10   REVIVING THE SECOND LAW ARGUMENT

In comparing the early uses of the second law argument to the manner in which it was used by young-Earth creationists, we see a striking decline in intellectual rigor. Clark and du Noüy were careful writers, and they thought seriously about their subject matter. We found much to criticize in their arguments, but they were not foolish in the way that Morris and his followers were.

However, in recent years anti-evolutionism has been dominated by the intelligent design (ID) movement, and one of their gambits has been to revive the second law argument. Unfortunately, their recent versions of the argument represent no improvement in intellectual seriousness over what we have seen.

The chief architect of this revival is mathematician Granville Sewell. His main claim is that the distinction between open and closed systems does little to mitigate the force of the second law argument. He has presented this view in a number of articles.

Let us consider some representative quotations. He writes:

> It is commonly argued that the spectacular increase in order which has occurred on Earth is consistent with the second law of thermodynamics because the earth is not an isolated system, and anything can happen in a non-isolated system as long as entropy increases outside the system compensate the entropy decreases inside the system.
>
> *(Sewell 2013a, 168)*

He later elaborates:

> [T]he whole idea of compensation, whether by distant or nearby events, makes no sense logically: an extremely improbable event is not rendered less improbable simply by the occurrence of "compensating" events elsewhere. According to this reasoning, the second law does not prevent scrap metal from reorganizing itself into a computer in one room, as long as two computers in the next room are rusting into scrap metal – and the door is open.
>
> *(Sewell 2013a, 170)*

Sewell often uses hypotheticals like this to mock the idea that entropy decreases in one part of a system can be compensated for by increases elsewhere. Here is another example:

I was discussing the second law argument with a friend recently, and mentioned that the second law has been called the "common sense law of physics." The next morning he wrote:

> Yesterday I spoke with my wife about these questions. She immediately grasped that chaos results in the long term if she would stop caring for her home.

I replied:

> Tell your wife she has made a perfectly valid application of the second law of thermodynamics. In fact, let's take her application a bit further. Suppose you and your wife go for a vacation, leaving a dog, a cat, and a parakeet loose in the house (I put the animals there to cause the entropy to increase more rapidly, otherwise you might have to take a much longer vacation to see the same effect). When you come back, you will not be surprised to see chaos in the house. But tell her some scientists say, "but if you leave the door open while on vacation, your house becomes an open system, and the second law does not apply to open systems … you may find everything in better condition than when you left." I'll bet she will say, "If a maid enters through the door and cleans the house, maybe, but if all that enters is sunlight, wind and other animals, probably not."

> *(Sewell 2013a, 175–176, ellipsis in original)*

Keep in mind that this quotation appeared in what was intended to be a scholarly article, as opposed to an article directed towards lay audiences. It was published as part of the proceedings of the academic conference on biological information I mentioned in Section 6.3. I take that to mean that Sewell, as well as the team of editors who assembled the proceedings volume, are happy to have this considered as a prime example of ID scholarship.

We will come to Sewell's discussion of the "compensation" argument momentarily, but first we must consider some of the sillier aspects of his writing, if only to appreciate, once again, why scientists become very impatient when responding to ID folks. In light of our previous discussion, Sewell's remarks are readily seen to have no connection to reality.

No one has ever claimed that "anything can happen" in an open system. The actual claim is that the second law is a precise mathematical statement. If your proposed physical process does not violate equation 7.2, then it is not ruled out by the second law. There might be reasons having nothing to do with thermodynamics for thinking the process is impossible, but you are, at least, good to go as far as the second law is concerned. Chemist Bob Lloyd, in a response to Sewell, expressed the point well:

> [W]e should note that the phrase "anything can happen" is highly tendentious. If one shows that a particular process is not forbidden by the Second Law, that falls far short of showing that it *can* happen.   *(Lloyd 2012, 30)*

It is very faint praise to say of a physical process that it is consistent with the second law, and this helps us to analyze Sewell's example of scrap metal turning into computers. This scenario is so far removed from the sorts of systems to which thermodynamics applies that it is unclear how to analyze it in terms relevant to the second law. If Sewell wants to provide a reversible process through which scrap metal assembles itself into computers, then we can revisit the question at that time. The statistical understanding of entropy is likewise unhelpful here. With energy entering the system, statistical mechanics has nothing to say about the forms into which matter will rearrange itself. Of course, we can reply in the same way to the example of a room cleaning itself.

The point is that our judgments about what will or will not happen when energy is added to a system have very little to do with abstract principles of thermodynamics. Instead, we have to make those judgments on a case by case basis depending on the specific nature of the system under consideration. It is obvious that sunlight will not cause scrap metal to assemble itself into computers, not because of anything we know from thermodynamics, but because we know that nothing much happens, either chemically or mechanically, when sunlight shines on metal. On the other hand, we might consider it equally obvious that sunlight *will* fuel the chemical reactions that allow a plant to grow and thrive, again not because of thermodynamics, but because we understand a great deal about biochemistry and plant physiology.

Let us now consider what Sewell says about the "compensation" argument. His assertion that "an extremely improbable event is not rendered less improbable simply by the occurrence of compensating events elsewhere," is difficult to understand. Surely it depends on the nature of the events in question. The chemical reactions that drive photosynthesis are unlikely to happen if a plant is enclosed in an opaque box. If that box is suddenly open to the sun, then those reactions become very probable.

Simple examples like this make it hard to understand precisely what Sewell thinks is "illogical" about the notion of entropy decreases in one part of a system being compensated for by entropy increases elsewhere. After all, what Sewell derides as "the compensation argument" is, for physicists, just an immediate and straightforward consequence of what the second law asserts.

The basis for Sewell's complaint about compensation appears to be his conviction that the entropy referred to in the second law can be usefully partitioned into many different types of entropy. He writes:

> [I]f we define "X-entropy" to be the entropy associated with any diffusing component X (for example, X might be heat), and, since entropy measures disorder, "X-order" to be the negative of X-entropy, a closer look at the equations for entropy change shows that they not only say that the X-order cannot increase in an isolated system, but that they also say that in a non-isolated system the X-order cannot increase faster than it is imported through the boundary.   (Sewell 2013a, 168)

This is hard to parse, and his maddeningly vague writing throughout the paper does little to clarify matters, but his basic point appears to be something like this: The thermal entropy referenced in the second law is just one instance of a more general phenomenon. There are actually many types of entropy, and it does not make sense to say that a decrease in one sort of entropy can be compensated for by an increase in some other sort of entropy. The thermal energy imported from the sun does nothing to help explain how the entropy decrease due to evolution is possible.

Understood in this way, the argument has two parts:

1. The entropy involved in the second law can be usefully partitioned into different types of entropy, and decreases in one type of entropy cannot be compensated for by increases in other types of entropy.
2. Evolutionary theory really does contradict the second law because an influx of "thermal order" cannot compensate for the entropy decrease due to evolution.

Now, several times in this book I have suggested that certain bold claims made by anti-evolutionists ought to trigger whatever skeptical impulses you possess. This is another of those times. The physics underlying the second law has been well understood for close to two centuries, and it is laid out with clarity and mathematical precision in any of the numerous textbooks on the subject. If someone tells you they have discovered a heretofore unnoticed aspect of the second law, and further claims that they can explain it in a few pages of mostly nontechnical writing, then you should suspect not that you are witnessing a revolution in physics, but simply that this person is in some way confused.

Let us start with the first point. The entropy referred to in the second law is a precisely defined quantity – essentially it is heat divided by temperature. The mathematical derivation of the second law is based on this definition. You are welcome to analyze other components of a system in terms that are reminiscent of how the second law treats entropy, you can make whatever definitions you want, and you are free to refer to things other than heat divided by temperature as an entropy of some sort. But in that case, whatever you are doing has nothing to do with the second law. Bob Lloyd, in the aforementioned discussion of Sewell's argument, makes the salient point. Referring to a specific example in which Sewell defined something called the "carbon entropy," Lloyd writes:

> The assumption that the various components of the total entropy can be treated independently has been imposed arbitrarily, without justification. In a block of steel the "thermal entropy" and the "carbon entropy" interact to give one observable quantity, the entropy; similar statements apply to any system. For any particular system in a specified condition it may be possible to define nominally separate *contributions* to the total entropy, but this does not make these contributions *independent* of each other. Such a conceptual separation is sometimes made for convenience, but it is essential to show that it is valid in any particular case.
>
> *(Lloyd 2012, 31)*

This leads nicely into the second point. If Sewell wants to apply his ideas to the evolutionary process. Then he needs to explain precisely which part of the process is ruled out by this invocation of thermodynamics. It is not sufficient to just deride compensation as illogical, or to assert that human evolution is impossible.

When scientists say that the thermal energy from the sun drives the evolutionary process, there is nothing in any way speculative about the mechanisms involved. We understand perfectly, at the level of individual atoms and molecules, how sunlight causes photosynthesis. We understand the full chain of causation for how sunlight ultimately provides the energy for all the processes of life. Moreover, as we have noted, every individual part of the evolutionary process is a well-understood, empirical fact: genes really do mutate, this sometimes leads to new functionalities, and natural selection really can string together these novelties into adaptive change. So, again, which part of this process is supposed to be ruled out by the second law?

We have shown that neither part of Sewell's argument is correct. It is not legitimate to partition the entropy in the manner he describes, and there is no plausible connection between his ideas and evolution. However, there is another aspect of Sewell's writing that merits some consideration.

Specifically, he asserts that a more expansive understanding of the second law would expose the conflict with evolutionary theory. He writes:

> Although all current statements of the second law apply only to isolated systems, the principle underlying the second law can actually be stated in a way that applies to open systems. ... [I]f the object is no longer isolated, then the thermal entropy can decrease, but no faster than it is exported. Stated another way, the thermal order (defined as the negative of thermal entropy) can increase, but no faster than it is imported. *(Sewell 2013b, 4)*

This leads to his own statement of the second law, "Natural (unintelligent) forces do not do macroscopically describable things that are extremely improbable from the microscopic point of view." (Sewell 2013b, 5)

He now completes his argument with this:

> The second law is all about using probability at the microscopic level to predict macroscopic change. The confusion in applying it to less quantifiable applications such as evolution is the result of trying to base it on something else, such as "entropy cannot decrease," when entropy may be difficult or impossible to define.

> This statement of the second law, or at least the fundamental principle behind the second law, is the one that should be applied to evolution. Those wanting to claim that the basic principle behind the second law is not violated ... need to argue that, under the right conditions, macroscopically describable things such as the spontaneous rearrangement of atoms into machines capable of mathematical computations, or of long-distance air travel, or of receiving pictures and sounds transmitted from the other side of

> the planet, or of interplanetary space travel, are not really
> astronomically improbable from the microscopic point of view,
> thanks to the influx of solar energy and to natural selection or
> whatever theory they use to explain the evolution of life and of
> human intelligence.   *(Sewell 2013b, 5)*

Starting at the beginning, it is not at all true that "all current statements of the second law apply only to isolated systems." We have noted that the most general possible statement of the second law is found in equation 7.2, and that equation is valid in any thermodynamical system. This equation, and the reasoning underlying it, can be found in any thermodynamics textbook.

Moreover, the notion that in an open system entropy can decrease, "but no faster than it is exported," is a very crude way of expressing what the equation says precisely. The equation tells us exactly the extent to which entropy can decrease in an open system.

The next claim is that the second law "is all about using probability at the microscopic level to predict macroscopic change." There is certainly some truth to this, as we discussed in Section 7.4. The problem comes when we attempt to take his advice about applying this understanding of the second law to evolutionary theory. In trying to do so, we immediately confront the issue we discussed at length in Chapter 5. Specifically, we have no way of determining the appropriate probability distribution.

Statistical mechanics finds its most natural applications in the theory of gases. When we have large ensembles of gas molecules, it is appropriate to apply a uniform distribution to the various microstates, and that justifies our conclusions about which macrostates are the most likely to be observed. There are other sorts of relatively simple physical systems whose macroscopic behavior can be usefully analyzed by considering the statistical behavior of large ensembles of smaller particles, and the methods of statistical mechanics will find applications to those systems as well.

This is not at all the situation with evolutionary theory. Imagine some moment in natural history at which large numbers of animals exist, and suppose we want to draw a conclusion about the future course of evolution. There is no insight to be gained by treating those animals as large ensembles of smaller particles. We have no basis at all for relating the probabilistic behavior of the particles to the macroscopic properties of the biosphere. The influx of solar energy ultimately fuels all of the physical processes on which life relies, and once life is present, it has a high probability of perpetuating itself into the future. As in Chapter 5, there is no way of quantifying these probabilities, and that is why biologists do not use probability theory to model the long-term trajectory of evolution.

Sewell's argumentation is striking for its level of abstraction. He speaks casually of improbability and the second law, but he never tells us precisely which part of the standard scientific account must be discarded. In the case of evolution, what are the macroscopically describable things he thinks are extremely improbable from the microscopic point of view?

Strangely, Sewell finds his examples not from evolution, but from the history of human technology. He is skeptical that atoms will spontaneously rearrange themselves into airplanes and whatnot. Now, we can certainly agree that on a lifeless planet, no amount of solar energy will directly fuel any process that will cause atoms to rearrange themselves into functional machines. You need intelligence for that. It is possible that after many millennia, animals with sufficient intelligence to build machines will appear, and then you might have airplanes and the rest. Or it might be that such intelligent animals never appear, and then you will not have those machines.

Regardless, the fact remains that once life appears and the evolutionary process gets started, arrangements of matter that would otherwise have been very improbable suddenly become not improbable at all.

Finally, we should note there is no such thing as the "fundamental principle behind the second law." The second law says what it says, and no more. In a legal setting, it is common to contrast the letter of the law with the spirit of the law, but there is no such distinction in science. In the end, therefore, the principles of thermodynamics do not actually play any role in Sewell's argument. He is only asserting his own incredulity over the theory of evolution and then using thermo-jargon to give those vague doubts a patina of scientific legitimacy.

This brings us to the end of our discussion of thermodynamics, and of the many abuses of the subject found in the literature of anti-evolutionism. Let us give the last word to Isaac Asimov, who aptly summarized the situation:

> The second law of thermodynamics (expressed in kindergarten terms) states that all spontaneous change is in the direction of increasing disorder, that is, in a "downhill" direction. There can be no spontaneous build-up of the complex from the simple, therefore, for that would be moving "uphill." Clearly, then, the creationist argument runs, since, by the evolutionary process, complex forms of life form from simple forms, that process, as described by scientists, defies the second law, and so creationism must be true.
>
> This sort of argument implies that a fallacy clearly visible to anyone is somehow invisible to scientists, who must therefore be flying in the face of the second law through sheer perversity. Scientists, however, *do* know about the second law and they are *not* blind. It's just that an argument based on kindergarten terms, as so many of the creationist arguments are, is suitable only for kindergartens.
>
> *(Asimov 1997, 9)*

7.11   NOTES AND FURTHER READING

There are many excellent books that discuss the basics of thermodynamics. For a readable, nontechnical introduction to the subject, I recommend the book by Goldstein and Goldstein (1993). For a more advanced, but still nontechnical introduction, I recommend the two books by Atkins (1984, 2010). For the reader with some familiarity with calculus who does not mind a few equations, but who would also prefer not to deal with the oppressive detail of the standard textbooks, I recommend the book by Van Ness (1969), which remains in print to this day. Finally, the two books by Lemons (2009, 2019) have many insightful things to say.

For a thorough discussion of the relation of thermodynamics to evolution, the classic book by Blum (1951) remains timely. A more recent discussion of the same topics can be found in the book by Ho (1997).

We have noted that entropy as physicists understand it does not really have much to do with everyday notions of order and disorder. However, it is common in nontechnical introductions to thermodynamics for writers to illustrate entropy with everyday examples like messy rooms or shuffled decks of cards. As we have noted, this is useful for communicating the flavor of the second law to nonscientific audiences, but it is not sufficient if you are engaging in serious scientific discourse. The two articles by Lambert (1999, 2002) and the article by Leff (2012) provide useful discussions about how easy it is to be misled if you think of entropy as synonymous with disorder.

In Section 7.8, I discussed a calculation carried out by Robert Oerter to estimate whether the entropy decrease due to evolution was in danger of violating the second law. His calculation was carried out using the classical understanding of entropy. Attempts to apply the statistical mechanical understanding of entropy to evolution have been attempted by physicists Daniel Styer (2008) and Emory Bunn (2009). Both concluded that evolution is safe by many orders of magnitude from any charge of violating the second law. Given the difficulties inherent in imagining a biosphere as a large ensemble of smaller particles, all such statistical attempts should be regarded as crude estimates. However, they do show, again, that the burden is very much on the anti-evolutionists to provide a serious justification for their use of the second law argument.

Also in Section 7.8, I discussed why it is problematic to apply the principles of classical thermodynamics directly to living organisms. In particular, classical thermodynamics deals with systems that are close to equilibrium, while living organisms are far from equilibrium. This has led to the development of "nonequilibrium thermodynamics," which is a major area of research among physicists today. Readable introductions to this branch of physics are not so easy to come by, but the book by Schneider and Sagan (2005) is a possible place to start. Some of their claims are controversial, but the early chapters of the book are a decent introduction to a difficult area. In a previous article, I made some brief remarks about nonequilibrium thermodynamics, as well as about connections between thermodynamics and information theory (Rosenhouse 2017).

I have not discussed the origin of life in this book, as I regard it as an entirely separate topic from the modern theory of biological evolution. However, for a recent discussion of the relationship of thermodynamics to the origin of life question, have a look at the book by England (2020).

8 Epilogue

8.1 AGAIN, BAD MATH CAN BE RHETORICALLY EFFECTIVE

We have completed our survey of mathematical anti-evolutionism, and we have found that none of its arguments is at all successful. However, the situation is actually even worse than that. There is one additional point to be made, which is this: When we consider the project of mathematical anti-evolutionism as a whole, we see that mathematics only plays a rhetorical role in its discourse.

For example, consider the notion of "complex, specified information (CSI)." As we saw in Chapter 5, ID proponents assert that they can use this concept to provide mathematical proof of design in natural history. Recall that an object was said to be complex and specified when it was highly improbable and also fit a recognizable pattern. The claim was made that adaptations like the bacterial flagellum are instances of CSI and therefore could not be produced solely by standard evolutionary mechanisms. This argument has been developed in particular by William Dembski.

But what did we see when he tried to apply his concept to a practical case? Specifically, what happened when he tried to show the flagellum is both complex and specified?

We saw, first, that his probability calculation was completely parasitic on the notion of "irreducible complexity," which we discussed in Section 2.5. Recall that this concept was developed by ID proponent Michael Behe. His claim was that if a complex biological system requires multiple, interdependent parts to function, then it could not have arisen through gradual evolution. Dembski explicitly based his calculation on the assumption that this claim was correct.

Now, if Behe's claim was correct, then that would be a strong argument all by itself that evolutionary theory had been falsified. A probability calculation would in no way make it stronger. We should note, after all, that Behe himself did not think his argument needed to be supplemented with any mathematics. However, since Behe's claim is entirely incorrect, for the reasons we discussed in Section 2.5, any calculation based on it can be dismissed as worthless.

Either way, Demsbki's mathematical hand-waving contributed nothing to the argument since all of the conceptual work was being done by irreducible complexity. If this concept is correct then no calculation is needed, and if it is not correct then no calculation based on it will have any force.

What about specificity? Dembski's theoretical development of this concept essentially required graduate-level training in mathematics. He helped himself to copious amounts of notation, jargon, Greek letters, and equations. Anyone unaccustomed to wading through prose of this sort could easily come away thinking it represented work of depth and profundity just from the level of technical detail in its presentation.

However, when it came time to discuss the specificity of an actual biological system, the flagellum in this case, all of the technical minutiae went clean out the window. For all the use Dembski made of his elaborate theoretical musings, they might as well never have existed at all. He just declared it obvious that the flagellum was specified and quickly moved on to other dubious claims. At no point did he attempt to relate anything in reality to the numerous variables and parameters he included in his mathematical theorizing.

We also saw that Dembski modeled flagellum assembly as a three-stage process of origination, localization, and configuration. We noted that this model was biologically absurd, but we saw in Section 5.8 that Steinar Thorvaldsen and Ola Hössjer attempted to revive it in their 2020 paper. There we were presented with more notation, jargon, and Greek letters.

But then nothing happened. We might have hoped that the presentation of so dramatic an equation was a prelude to an equally dramatic insight into biology, but there was nothing of that sort. It turned out they did not introduce this formal equation with the intent of using it to calculate or prove anything. They just sort of dumped it on the page and then quickly moved on to the next thing. It was another instance of mathematics being used strictly for rhetorical effect.

In Chapter 6 we considered a new line of attack. We were told that cutting-edge mathematical results – the "No Free Lunch" theorems – exposed some sort of fallacy at the heart of evolutionary theory. ID proponents boasted of their own theoretical developments of these theorems, claiming to have extended them in ways that were directly relevant to assessing evolution's fundamental soundness. They pompously referred to their results as "conservation of information" theorems, thereby pretending that their trivial musings should be placed on the mantle next to the other great conservation principles of physics.

From there a familiar pattern played out. When we examined the biological conclusions said to be underwritten by these mathematical results, we found them to be all but vacuous. Despite draping their mathematics over many tedious, symbol-laden pages, it turned out the ID proponents just wanted to convince us that the environment has to embody certain physical principles for evolution to work. Most of us considered that obvious without any need for mathematics, but the ID proponents nonetheless presented it as some brilliant insight they discovered.

The pattern played out again in Chapter 7, when we considered the use of thermodynamics in anti-evolutionist discourse. In particular, we considered the claim that evolution conflicts with the second law. Now, thermodynamics is a very well researched branch of science. Its practitioners have developed tremendously sophisticated mathematical models for handling various sorts of systems, and these models have proven their worth through numerous practical successes in physics and engineering.

But anti-evolutionists do not use any of this in their discourse. The powerful mathematics of real research in thermodynamics may as well not exist for all the use made of it in their writing. Shorn of the physics jargon, their thermodynamical arguments amount to little more than blunt assertions of incredulity that evolution can increase the complexity of organisms over time. Once again, we see that powerful concepts from physics and mathematics are playing only a rhetorical role in their arguments. If you are not using any of the mathematical machinery of thermodynamics, then there is no reason to bring it up at all. Everyone agrees that complex biological systems require a special kind of explanation. Biologists rightly believe they have a sufficient explanation while anti-evolutionists demur, but the important point is that this discussion has nothing to do with thermodynamics.

This pattern, of introducing difficult mathematical concepts without ever really using them for any serious purpose, is ubiquitous in anti-evolution discourse, and this fact goes a long way to explaining why mathematicians and scientists are so disdainful of it. Professionals in these areas strive for the utmost clarity when presenting their work. Used properly, the jargon and notation permit a level of precision that simply cannot be achieved with more natural language. This might seem hard to believe, since a modern scientific research paper will be unreadable for anyone without significant training in the relevant discipline. But the problem is not a lack of clarity in the writing. Rather, it is just that the concepts involved are difficult, and experience is needed to become comfortable with them.

If you have the necessary training and experience, you quickly develop a nose for counterfeits. You can easily distinguish serious work in a discipline from propaganda and rhetoric. You know when jargon and notation are being used to speak with precision about a complex topic, and you know when it is just being used to bamboozle a lay audience.

In Section 2.6, I remarked that anti-evolutionist arguments play well in front of friendly audiences because in that environment the speakers never pay a price for being wrong. The response would be a lot chillier if they tried the same arguments in front of audiences with the relevant expertise. Try telling a roomful of mathematicians that you can refute evolutionary theory with a few back-of-the-envelope probability calculations, and see how far you get. Tell a roomful of physicists that the second law of thermodynamics conflicts with evolutionary theory, or a roomful of computer scientists that obscure theorems from combinatorial search have profound relevance to biology.

You will be lucky to make it ten minutes before the audience stops being polite.

8.2 CAN INTELLIGENCE BUILD COMPLEX ADAPTATIONS?

There is a final point to be made before wrapping this up.

Anti-evolutionists employ a bizarre double standard in assessing the relative merits of natural selection and intelligent design.

As we have seen, scientists point to copious physical evidence in support of their view that evolution by natural selection can craft complex structures. They point to the small-scale evidence of field studies of natural selection, and to the large-scale evidence of common descent. They note that all complex adaptations studied to date show clear signs of arising not from intelligent engineering, but through a long historical process. They point to numerous specific examples where we have strong evidence from paleontology, embryology, and genetics to tell us what the stepping-stones actually were.

Anti-evolutionists scoff at all of this. They dismiss it out of hand. They accuse scientists of making absurd extrapolations and of being blinded by materialist bias. They suggest that until you evolve a flagellum from scratch in a laboratory you have nothing.

Contrast that with how they treat intelligent design. For example, in his book *Undeniable,* ID proponent Douglas Axe writes:

> We're left to think that poor Tavros 2 [a solar-powered,
> underwater vehicle] is really no more worthy of comparison to a
> lowly cyanobacterium than it is to an exalted dolphin. After all,
> raw natural ingredients like sand and metal ores and crude oil
> become Tavros 2 only with the skillful help of thousands of people
> at hundreds of industrial plants of various kinds. With all due
> respect, this human invention does very little in comparison to
> the human effort expended to manufacture it. The contrast with
> cyanobacteria could hardly be more stark.   *(Axe 2016, 175)*

This sort of thing is commonplace throughout Axe's book. He is constantly telling us that the simplest biological processes are *way* beyond the puny contrivances of human engineers.

If that is true, then why should I not conclude that intelligence is fundamentally incapable of accomplishing what ID proponents attribute to it? If the greatest accomplishments of the greatest intelligences we know of are like nothing compared to the living world, then why the confidence that intelligence is responsible for the living world, much less for the universe as a whole?

Stephen Meyer has echoed Axe's arguments. In a section of his book *Darwin's Doubt* entitled "A Cause Now in Operation" he writes:

> Intelligent agents, due to their rationality and consciousness, have
> demonstrated the power to produce specified or functional
> information in the form of linear sequence-specific arrangements
> of characters. ... Our experience-based knowledge of information
> flow confirms that systems with large amount of specified or
> functional information invariably originate from an intelligent
> source.   *(Meyer 2013, 360)*

Intelligent agents can communicate with one another and build machines, but only on a very limited scale. Humans can do simple things like write books or build automobiles, but, as ID proponents are so keen to point out, building even the simplest microorganism, much less a horse or an eagle, already seems to be well beyond what intelligence can do. And from the ID perspective, this is just the tip of the iceberg. They say intelligence is capable of adjusting fundamental physical constants and of bringing whole worlds into being just with acts of will. This is all orders of magnitude beyond anything intelligence has ever been seen to do.

When scientists point to the copious circumstantial evidence that complex adaptations are the products of natural selection, ID proponents accuse them of making unwarranted extrapolations. But in inferring that the living world must be the product of intelligent design, they are guilty of a far more extravagant extrapolation than anything conceived of by modern biology. However it is that ID proponents arrive at their conclusions, they are certainly not extrapolating from "causes now in operation." Based on our experience, or on comparisons of human engineering to the natural world, the obvious conclusion is that intelligence cannot at all do what they claim it can do. Not even close.

Their argument is no better than saying that since moles are seen to make molehills, mountains must be evidence for giant moles.

8.3 CODA

It has been many years since I lived at home, and my father is now long retired. Recently, I reminded him of the conversation I described in Section 2.4. I mentioned that I now like to use that system of roads as an example of distinguishing systems that are designed from those that evolved.

He laughed. Apparently, in the years since I left home, that intersection has been completely redesigned. The powers that be got tired of cleaning up after the numerous crashes and human misery resulting from the poor design of the roads. So they shut it all down for several months and completely redid the whole thing. Now the arrangement of roads makes perfect sense, and the number of crashes there has declined dramatically.

The anti-evolutionists are right about one thing: we really can distinguish systems that were designed from those that evolved gradually. Unfortunately for them, the anatomy of organisms points overwhelmingly toward evolution and just as overwhelmingly away from design.

No piece of abstract mathematics is going to change that simple fact.

Bibliography

Applebaum, D. (1996). *Probability and Information: An Integrated Approach.* Cambridge: Cambridge University Press.

Asimov, I. (1997). The army of the night. In I. Asimov (ed.) *The Roving Mind.* Amherst: Prometheus Books.

Atkins, P. (1984). *The Second Law.* New York: Scientific American Books – W. H. Freeman.

Atkins, P. (2010). *The Laws of Thermodynamics: A Very Short Introduction.* New York: Oxford University Press.

Avise, J. C. (2010). *Inside the Human Genome: A Case for Non-Intelligent Design.* New York: Oxford University Press.

Axe, D. (2004). Estimating the prevalence of protein sequences adopting functional enzyme folds. *Journal of Molecular Biology.* 341(5): 1295–1315.

Axe, D. (2016). *Undeniable: How Biology Confirms Our Intuitions that Life Is Designed.* New York: HarperOne.

Barash, D. P. (2003). *The Survival Game: How Game Theory Explains the Biology of Cooperation and Competition.* New York: Times Books.

Beeby, M., Ferreira, J. L., Tripp, P., Albers, S.-V., and Mitchell, D. R. (2020). Propulsive nanomachines: The convergent evolution of archaella, flagella, and cilia. *FEMS Microbiology Reviews.* 44(3): 253–304.

Behe, M. J. (1996). *Darwin's Black Box: The Biochemical Challenge to Evolution.* New York: Free Press.

Behe, M. J. (2007). *The Edge of Evolution: The Search for the Limits of Darwinism.* New York: Free Press.

Behe, M. J. (2020). *A Mousetrap For Darwin: Michael J. Behe Answers his Critics.* Seattle: Discovery Institute Press.

Berkman, M. and Plutzer, E. (2010). *Evolution, Creationism, and the Battle to Control America's Classrooms.* Cambridge: Cambridge University Press.

Betts, E. H. (1944). Evolution and entropy. *Journal of the Transactions of the Victoria Institute.* 76: 1–27.

Blancke, S., Boudry, M., and Braeckman, J. (2011). Simulation of biological evolution under attack, but not really: A response to Meester. *Biology and Philosophy.* 26(1): 113–118.

Blum, H. (1951). *Time's Arrow and Evolution*. Princeton: Princeton University Press.

Bridgman, P. (1961). *The Nature of Thermodynamics*. 2nd ed. New York: Harper and Brothers.

Brillouin, L. (1949). Life, thermodynamics, and cybernetics. *American Scientist*. 37(4): 554–568.

Bunn, E. (2009). Evolution and the second law of thermodynamics. *American Journal of Physics*. 77(10): 922–925.

Burling, R. (1953). Evolution and thermodynamics. *Science Education*. 37(3): 184.

de Camp, L. Sprague (1968). *The Great Monkey Trial*. New York: Doubleday.

Carrier, R. C. (2004). The argument from biogenesis: Probabilities against a natural origin of life. *Biology and Philosophy*. 21(5): 739–764.

Carroll, S. B. (2007). God as genetic engineer. *Science*. 316(5380): 1427–1428.

Clark, R. E. D. (1943). Evolution and entropy. *Journal of the Transactions of the Victoria Institute*. 75: 49–71.

Conway Morris, S. (2003). *Life's Solution: Inevitable Humans in a Lonely Universe*. Cambridge: Cambridge University Press.

Coyne, J. (2009). *Why Evolution Is True*. New York: Viking.

Crosby, J. L. (1967). Computers in the study of evolution. *Science Progress*. 55(218): 279–292.

Darwin, C. (1859). *On the Origin of Species*. New York: Gramercy Books, 1979 reprint edition.

Darwin, C. (1862). *On the Various Contrivances by Which British and Foreign Orchids Are Fertilised by Insects, and on the Good Effects of Intercrossing*. London: John Murray.

Dawkins, R. (1986). *The Blind Watchmaker: How the Evidence of Evolution Reveals a Universe Without Design*. New York: Norton.

Dawkins, R. (1996). *Climbing Mount Improbable*. New York: Norton.

Dawkins, R. (2003). *A Devil's Chaplain: Reflections on Hope, Lies, Science, and Love*. New York: Houghton Mifflin.

Dawkins, R. (2009). *The Greatest Show on Earth: The Evidence for Evolution*. New York: Free Press.

Dembski, W. A. (1998). *The Design Inference: Eliminating Chance Through Small Probabilities*. Cambridge: Cambridge University Press.

Dembski, W. A. (1999). *Intelligent Design: The Bridge Between Science and Theology*. Downers Grove: InterVarsity Press.

Dembski, W. A. (2002). *No Free Lunch: Why Specified Complexity Cannot be Purchased Without Intelligence*. Lanham: Rowman & Littlefield.

Dembski, W. A. (2004). *The Design Revolution: Answering the Toughest Questions About Intelligent Design*. Downers Grove: InterVarsity Press.

Dembski, W. A. (2005). Specification: The pattern that signifies intelligence. Online at https://billdembski.com/documents/2005.06.Specification.pdf. Last accessed December 2021.

Divine, S. (2014). An algorithmic information challenge to intelligent design. *Zygon*. 49(1): 42–65.

Dobzhansky, T. (1973). Nothing in biology makes sense except in the light of evolution. *The American Biology Teacher*. 35(3): 125–129.

Dronamraju, K. (2011). *Haldane, Mayr, and Beanbag Genetics*. Oxford: Oxford University Press.

Dryden, T. F., Thomson, A. R., and White, J. H. (2008). How much of protein sequence space has been explored by life on Earth? *Journal of the Royal Society Interface*. 5(25): 953–956.

Dugatkin, L. A. and Reeve, H. K. eds. (1998). *Game Theory and Animal Behavior*. New York: Oxford University Press.

Eddington, A. (1929). *The Nature of the Physical World*. New York: The MacMillan Company.

Elsberry, W. and Shallit, J. (2011). Information theory, evolutionary computation, and Dembski's "complex specified information." *Synthese*. 178(2): 237–270.

England, J. (2020). *Every Life Is on Fire: How Thermodynamics Explains the Origins of Living Things*. New York: Basic Books.

English, T. (2017). Evo-info review: Do not buy the book until .... Online at www.theskepticalzone.com/wp/evo-info-review-do-not-buy-the-book-until/ Last accessed December 2021.

Ferrada, E. and Wagner, A. (2010). Evolutionary innovations and the organization of protein functions in genotype space. *PLoS ONE*. 5(11): 1–11.

Fisher, R. A. (1935). *The Design of Experiments*. Edinburgh: Oliver and Boyd.

Fitelson, B., Stephens, C., and Sober, E. (1999). How not to detect design—Critical notice: William A. Dembski, *The Design Inference*. *Philosophy of Science*. 66(3): 472–488.

Fogel, L. J. (1999). *Intelligence Through Simulated Evolution: Forty Years of Evolutionary Programming*. New York: John Wiley and Sons.

Forbes, N. (2004). *Imitation of Life: How Biology Is Inspiring Computing*. Cambridge: Cambridge University Press.

Forrest, B. and Gross, P. R. (2003). *Creationism's Trojan Horse: The Wedge of Intelligent Design*. New York: Oxford University Press.

Foster, D. (1999). Proving God exists. *The Saturday Evening Post*. November–December: 59–61, 78, 80–81, 84.

Gavrilets, S. (2010). High-dimensional fitness landscapes and speciation. In: *Evolution: The Extended Synthesis*. M. Pigliucci and G. Mueller, eds. Cambridge: MIT Press. 45–80.

Gillespie, J. H. (2004). *Population Genetics: A Concise Guide*. 2nd ed. Baltimore: Johns Hopkins University Press.

Gishlick, A. (2004). Evolutionary paths to irreducible complexity: The avian flight apparatus. In: *Why Intelligent Design Fails: A Scientific Critique of the New Creationism*. M. Young, T. Edis, eds. 58–71.

Gitt, W. (2001). *In the Beginning Was Information*. Germany: Christliche Literatur-Verbreitung.

Gitt, W., Compton, R., and Fernandez, J. (2011). Biological information – What is it? In: *Biological Information: New Perspectives*. R. J. Marks II, M. J. Behe, W. A. Dembski, B. L. Gordon, J. C. Sanford, eds. Singapore: World Scientific. 11–25.

Glaeser, G. and Paulus, H. F. (2015). *The Evolution of the Eye*. Switzerland: Springer International Publishing.

Godfrey-Smith, P. (2001). Information and the argument from design. In: *Intelligent Design Creationism and Its Critics*. R. T. Pennock, ed. Cambridge: MIT Press. 575–596.

Goldstein, M. and Goldstein, I. (1993). *The Refrigerator and the Universe*. Cambridge: Harvard University Press.

Gould, S. J. (1980). *The Panda's Thumb: More Reflections in Natural History*. New York: Norton.

Gould, S. J. (1988). *Wonderful Life: The Burgess Shale and the Nature of History.* New York: Norton.

Gould, S. J. (1993). *Eight Little Piggies: Reflections in Natural History.* New York: Norton.

Gowers, T. (2002). *Mathematics: A Very Short Introduction.* Oxford: Oxford University Press.

Gray, A. (1876). *Darwiniana: Essays and Reviews Pertaining to Darwinism*. New York: D. Appleton and Co.

Gregory, T. R. (2008). The evolution of complex organs. *Evolution: Education and Outreach*. 1(4): 358–389.

Griffiths, P. (2001). Genetic information: A metaphor in search of a theory. *Philosophy of Science*. 68(3): 394–412.

Hafer, A. (2015). *The Not-So-Intelligent Designer: Why Evolution Explains the Human Body and Intelligent Design Does Not*. Eugene: Cascade Books.

Häggström, O. (2007). Intelligent design and the NFL theorems. *Biology and Philosophy*. 22(2): 217–230.

Harper, J. L. (1968). Evolution: What is required of a theory? *Science*. 160(3826): 408.

Hazen, R. M., Griffin, P. L., Carothers, J. M., and Szostak, J. W. (2007). Functional information and the emergence of biocomplexity. *Proceedings of the National Academy of Sciences*. 104(supp. 1): 8574–8581.

Ho, M. (1997). *The Rainbow and the Worm: The Physics of Organisms*. Singapore: World Scientific.

Hopkins, B. and Wilson, R. (2004). The truth about Königsberg. *College Mathematics Journal*. 35(3): 198–207.

Hunt, A. (2007). Axe (2004) and the evolution of enzyme function. Online at: https://pandasthumb.org/archives/2007/01/92-second-st-fa.html. Last accessed February 2021.

Huxley, J. (1942). *Evolution: The Modern Synthesis*. London: Allen & Unwin.

Isaac, R. (2017). Review of *Introduction to Evolutionary Informatics* by R. J. Marks II, W. A. Dembski, and W. Ewert. *Perspectives on Science and Christian Faith*. 69(2): 99–104.

Isaak, M. (2007). *The Counter-Creationism Handbook*. Berkeley: University of California Press.

Johnson, P. (1991). *Darwin on Trial*. Washington, DC: Regnery Gateway.

Johnson, P. (2000). *The Wedge of Truth: Splitting the Foundations of Naturalism*. Downers Grove: InterVarsity Press.

Kimura, M. (1961). Natural selection as the process of accumulating genetic information in adaptive evolution. *Genetics Research*. 2(1): 127–140.

Kitcher, P. (2007). *Living With Darwin: Evolution, Design, and the Future of Faith*. New York: Oxford University Press.

Lambert, F. (1999). Shuffled cards, messy desks, and disorderly dorm rooms— Examples of entropy increase? Nonsense! *Journal of Chemical Education*. 76(10): 1385–1387.

Lambert, F. (2002). Disorder—A cracked crutch for supporting entropy discussions. *Journal of Chemical Education*. 79(2): 187–192.

Land, M. and Nilsson, D-E. (2012). *Animal Eyes*. New York: Oxford University Press.

Lane, N. (2015). *The Vital Question: Energy, Evolution, and the Origins of Complex Life*. New York: W. W. Norton & Company.

Larson, E. (2020). *Summer for the Gods: The Scopes Trial and America's Continuing Debate Over Science and Religion*. (Reprint of the original 2000 edition with a new afterword.) New York: Hachette.

Lecomte du Noüy, P. (1947). *Human Destiny*. New York: Longmans, Green and Co.

Leff, H. S. (2012). Removing the mystery of entropy and thermodynamics—Part V. *Physics Teacher*. 80(5): 274–276.

Lehman, J., Clune, J., Misevic D., et al. (2020). The surprising creativity of digital evolution: A collection of anecdotes from the evolutionary computation and artificial life research communities. *Artificial Life*. 26(2): 274–306.

Lemons, D. (2009). *Mere Thermodynamics*. Baltimore: Johns Hopkins University Press.

Lemons, D. (2019). *Thermodynamic Weirdness: From Fahrenheit to Clausius*. Cambridge: MIT Press.

Lenski, R. E., Ofria, C., Pennock, R. T., and Adami, C. (2003). The evolutionary origin of complex features. *Nature*. 423(6936): 139–144.

Lents, N. H. (2018). *Human Errors: A Panaroma of Our Glitches, From Pointless Bones to Broken Genes*. Boston: Mariner Books.

Lloyd, B. (2012). Is there any conflict between evolution and the second law of thermodynamics? *The Mathematical Intelligencer*. 34(1): 29–33.

Losos, J. B. (2011). *In the Light of Evolution: Essays from the Laboratory and Field*. Greenwood Village: Roberts and Co.

Losos, J. B. (2018). *Improbable Destinies: Fate, Chance, and the Future of Evolution*. New York: Riverhead Books.

Losos, J. B. and Lenski, R. E. (2016). *How Evolution Shapes Our Lives: Essays on Biology and Society*. Princeton: Princeton University Press.

Lynch, M. (2005). Simple evolutionary pathways to complex proteins. *Protein Science*. 14(9): 2217–2225.

Marks II, R. J., Behe, M. J., Dembski, W. A., Gordon, B. L., and Sanford, J. C. eds. (2013). *Biological Information: New Perspectives*. Singapore: World Scientific.

Marks II, R. J., Dembski, W. A., and Ewert, W. (2017). *Introduction to Evolutionary Informatics*. Singapore: World Scientific.

Matzke, N. (2007). The edge of creationism. *Trends in Ecology and Evolution*. 22(11): 566–567.

Matzke, N. (2009). But isn't it creationism? In: *But Is it Science? The Philosophical Question in the Evolution/Creation Controversy*. R. Pennock and M. Ruse, eds. 377–413. Amherst: Prometheus Books.

Mayfield, J. E. (2013). *The Engine of Complexity: Evolution as Computation*. New York: Columbia University Press.

Maynard Smith, J. (1982). *Evolution and the Theory of Games*. Cambridge: Cambridge University Press.

Maynard Smith, J. (1992). Byte-sized evolution. *Nature*. 355(6363): 772–773.

McIntosh, A. C. (2009). Information and entropy—top-down or bottom-up development in living systems? *International Journal of Design & Nature Ecodynamics*. 4(4): 351–385.

McLennan, D. A. (2008). The concept of co-option: Why evolution often looks miraculous. *Evolution: Education and Outreach*. 1(3): 247–258.

McNamara, J. M. and Leimer, O. (2020). *Game Theory in Biology: Concepts and Frontiers*. New York: Oxford University Press.

McNickle, G. G. and Dybzinski, R. (2013). Game theory and plant ecology. *Ecology Letters*. 16(4): 545–555.

Meester, R. (2009). Simulation of biological evolution and the NFL theorems. *Biology and Philosophy*. 24(4): 461–472.

Meyer, S. (2013). *Darwin's Doubt*. New York: HarperOne.

Meyer, S. (2017). Neo-Darwinism and the origin of biological form and information. In: *Theistic Evolution: A Scientific, Philosophical, and Theological Critique*, J. P. Moreland, S. C. Meyer, C. Shaw, A. K. Gauger, and W. Grudem, eds. Wheaton: Crossway.

Miller, K. (2007). Falling over the edge. *Nature*. 447(7148): 1055–1056.

Moorhead, P. S. and Kaplan, M. M., eds. (1967). *Mathematical Challenges to the Neo-Darwinian Interpretation of Evolution*. Philadelphia: Wistar Institute Press.

Moran, L. (2020). Of mice and Michael. Online at: https://sandwalk.blogspot.com/2020/12/of-mice-and-michael.html. Accessed December 2021.

Morris, H. M. and Parker, G. E. (1987). *What Is Creation Science?* (Revised and Expanded Edition). El Cajon: Master Books.

Morris, H. M. (1982). *The Troubled Waters of Evolution*, 2nd ed. San Diego: Creation-Life Publishers.

Muller, H. (1918). Genetic variability, twin hybrids and constant hybrids in a case of balanced lethal factors. *Genetics*. 3(5): 422–499.

Musgrave, I. (2004). Evolution of the bacterial flagellum. In: *Why Intelligent Design Fails: A Scientific Critique of the New Creationism*. M. Young and T. Edis, eds. pp. 72–84.

Nowak, M. A. (2006). *Evolutionary Dynamics: Exploring the Equations of Life*. Cambridge: Belknap/Harvard.

Numbers, R. (2006). *The Creationists: From Scientific Creationism to Intelligent Design*. Expanded ed. Cambridge: Harvard University Press.

Oerter, R. (2006). Does life on earth violate the second law of thermodynamics? Online at http://physics.gmu.edu/~roerter/EvolutionEntropy.htm Accessed December 2021.

Olofsson, P. (2008). Intelligent design and mathematical statistics: A troubled alliance. *Biology and Philosophy*. 23(4): 545–553.

Olofsson, P. (2013). Probability, statistics, evolution, and intelligent design. *Chance*. 21(3): 42–45.

Orgel, L. (1973). *The Origins of Life: Molecules and Natural Selection.* New York: John Wiley and Sons.

Paley, W. (2006). *Natural Theology.* Oxford World Classics reprint edition. New York: Oxford University Press.

Pierce, J. R. (1961). *An Introduction to Information Theory: Symbols, Signals, and Noise.* Mineola: Dover.

Pieret, J. (2006). The Quote Mine Project, or Lies, Damned Lies and Quote Mines. Online at www.talkorigins.org/faqs/quotes/mine/project.html. Last accessed December 2021.

Pigliucci, M. and Müller, G. (2010), eds. *Evolution: The Extended Synthesis.* Cambridge: The MIT Press.

Prothero, D. R. (2007). *Evolution: What the Fossils Say and Why it Matters.* New York: Columbia University Press.

Prothero, D. R. (2020). *The Story of Evolution in 25 Discoveries: The Evidence and the People Who Found it.* New York: Columbia University Press.

Purdom, G. (2008). Evidence of new genetic information? Online at https://answersingenesis.org/genetics/mutations/evidence-of-new-genetic-information. Last accessed December 2021.

Rogers, A. R. (2011). *The Evidence for Evolution.* Chicago: The University of Chicago Press.

Rosenhouse, J. (2002a). Rhetorical legerdemain in intelligent design literature. In *Darwin Day: Collection One,* A. Chesworth, S. Hill, K. Lipovsky, E. Snyder, and W. Chesworth, eds. Albuquerque: Tangled Bank Press. 327–338.

Rosenhouse, J. (2002b). Probability, information theory, and evolution. *Evolution.* 56(8): 1721–1722.

Rosenhouse, J. (2005). Why scientists get so angry when dealing with ID proponents. *Skeptical Inquirer.* 29(6): 42–45.

Rosenhouse, J. (2011). *Among the Creationists: Dispatches From the Anti-Evolutionist Front Line.* New York: Oxford University Press.

Rosenhouse, J. (2016). On mathematical anti-evolutionism. *Science and Education.* 25(1): 95–114.

Rosenhouse, J. (2017). Thermodynamical arguments against evolution. *Science and Education.* 26(1): 3–25.

Rosenhouse, J. (2018). Reviews of *Undeniable: How Biology Confirms that Life Is Designed* by D. Axe, *Introduction to Evolutionary Informatics,* by R. J. Marks II, W. A. Dembski, and W. Ewert, and *Proving Darwin: Making Biology Mathematical,* by G. Chaitin. *American Mathematical Monthly.* 125(6): 571–576.

Roth, A. A. (1998). *Origins: Linking Science and Scripture*. Silver Spring: Review and Herald Publishing Association.

Salsburg, D. (2001). *The Lady Tasting Tea: How Statistics Revolutionized Science in the Twentieth Century*. New York: Henry Holt and Co.

Sarkar, S. (2007). *Doubting Darwin?: Creationist Designs on Evolution*. Malden: Blackwell Publishing.

Schneider, E. and Sagan, D. (2005). *Into the Cool: Energy Flow, Thermodynamics, and Life*. Chicago: University of Chicago Press.

Schwab, I. R. (2012). *Evolution's Witness: How Eyes Evolved*. New York: Oxford University Press.

Scott, E. C. (2009). *Evolution vs. Creationism: An Introduction*. 2nd ed. Westport: Greenwood Press.

Sewell, G. (2013a). Entropy, evolution, and open systems. In: *Biological Information: New Perspectives*. R. J. Marks II, M. J. Behe, W. A. Dembski, B. L. Gordon, and J. C. Sanford, eds. Singapore: World Scientific. 168–178.

Sewell, G. (2013b). Entropy and evolution. *BIO-Complexity*. 3: 1–5.

Shannon, C. E. (1948). A mathematical theory of communication. *Bell System Technical Journal*. 27(3): 379–423.

Shubin, N. (2020). *Some Assembly Required: Decoding Four Billion Years of Life, from Ancient Fossils to DNA*. New York: Pantheon.

Sober, E. (2002). Intelligent design and probability reasoning. *International Journal for Philosophy of Religion*. 52(1): 65–80.

Stearns, S. C. (2020). Frontiers in molecular evolutionary medicine. *Journal of Molecular Evolution*. 88(1): 3–11.

Styer, D. (2008). Evolution and entropy. *American Journal of Physics*. 76(11): 1031–1033.

Szathmáry, E. and Maynard Smith, J. (1995). The major evolutionary transitions. *Nature*. 374(6519): 227–232.

Takagi, Y. A., Nguyen, D. H., Wexler, T. B., and Goldman, A. D. (2020). The coevolution of cellularity and metabolism following the origin of life. *Journal of Molecular Evolution*. 88(7): 598–617.

Taylor, J. (2015). *Body by Darwin: How Evolution Shapes Our Health and Transforms Medicine*. Chicago: University of Chicago Press.

Taylor, J. and Raes, J. (2004). Duplication and divergence: The evolution of new genes and old ideas. *Annual Review of Genetics*. 38: 615–643.

Theobald, D. (2012). 29+ evidences for macroevolution: The case for common descent. Online at www.talkorigins.org/faqs/comdesc/. Last accessed December 2021.

Thorvaldsen, S. and Hössjer, O. (2020). Using statistical methods to model the fine-tuning of molecular machines and systems. *Journal of Theoretical Biology*. 507: 1–14.

Ulam, S. (1972). Some ideas and prospects in biomathematics. *Annual Review of Biophysics and Bioengineering*. 1: 277–292.

Van Ness, H. C. (1969). *Understanding Thermodynamics*. New York: McGraw Hill.

Wein, R. (2002). Not a free lunch, but a box of chocolates. Online at: www.talkorigins.org/design/faqs/nfl/. Last accessed December 2021.

Whitcomb Jr., J. and Morris, H. (1961). *The Genesis Flood: The biblical Record and Its Scientific Implications*. Phillipsburg: Presbyterian and Reformed Publishing Company.

Williams, W. A. (1928). *The Evolution of Man Scientifically Disproved, in 50 Arguments*. Camden: W. A. Williams.

Wilson, R. (1996). *Introduction to Graph Theory*, 4th ed. Essex: Addison Wesley Longman.

Wolpert, D. H. (2002). William Dembski's treatment of the No Free Lunch theorems is written in jello. *Mathematical Reviews*, Feb. 2003. Available online at www.talkreason.org/articles/jello.cfm. Accessed December 2021.

Wolpert, D. H. and Macready, W. G. (1997). No free lunch theorems for optimization. *IEEE Transactions on Evolutionary Computation*. 1(1): 67–82.

Yeates, T. O. (2007). Protein structure: Evolutionary bridges to new folds. *Current Biology*. 17(2): R48–R50.

Index

Printed in the United States
by Baker & Taylor Publisher Services